S0-ACB-029

MORTALITY PATTERNS AND TRENDS IN THE UNITED STATES

Recent Titles in
Studies in Population and Urban Demography

Changing Sex Differential in Mortality
Robert D. Retherford

Teenage Marriages: A Demographic Analysis
John R. Weeks

Urbanization, Population Growth, and Economic Development in the Philippines
Ernesto del Mar Pernia

Success or Failure? Family Planning Programs in the Third World
Donald J. Hernandez

Soviet Jewry Since the Second World War:
Population and Social Structure
Mordechai Altshuler

Mortality of Hispanic Populations: Mexicans, Puerto Ricans, and Cubans in the United States and in the Home Countries
Ira Rosenwaike, editor

MORTALITY PATTERNS AND TRENDS IN THE UNITED STATES

Paul E. Zopf, Jr.

Studies in Population and Urban
Demography, Number 7

GREENWOOD PRESS
Westport, Connecticut • London

HB
1335
Z67
1992

FLORIDA STATE
UNIVERSITY LIBRARIES

JUL 12 1993

TALLAHASSEE, FLORIDA

Library of Congress Cataloging-in-Publication Data

Zopf, Paul E.
    Mortality patterns and trends in the United States / Paul E. Zopf,
Jr.
      p. cm. — (Studies in population and urban demography, ISSN
0147–1104 ; no. 7)
    Includes bibliographical references and index.
    ISBN 0–313–26769–3 (alk. paper)
    1. Mortality—United States. I. Title. II. Series.
HB1335.Z67 1992
304.6′4′0973—dc20   92–15488

British Library Cataloguing in Publication Data is available.

Copyright © 1992 by Paul E. Zopf, Jr.

All rights reserved. No portion of this book may be
reproduced, by any process or technique, without the
express written consent of the publisher.

Library of Congress Catalog Card Number: 92–15488
ISBN: 0–313–26769–3
ISSN: 0147–1104

First published in 1992

Greenwood Press, 88 Post Road West, Westport, CT 06881
An imprint of Greenwood Publishing Group, Inc.

Printed in the United States of America

The paper used in this book complies with the
Permanent Paper Standard issued by the National
Information Standards Organization (Z39.48–1984).

10 9 8 7 6 5 4 3 2 1

TO
JAMES C. BREWER,
HEALER AND FRIEND
AND
VERNIE L. DAVIS,
COLLEAGUE AND FRIEND

Contents

Figures

Tables

Preface

Dramatic ongoing changes in American health care and its availability to various groups in the population necessitate continued updating of information and interpretation of the nation's mortality patterns and trends. We need to know the realities of general mortality, differentials by various demographic characteristics, infant mortality, cause of death, and life expectancy, and to understand how they are changing. We also need to understand how social, economic, political, and other conditions affect mortality patterns and trends, and, in turn, how those patterns and trends affect social, economic, political, and other conditions. Thus, the mortality analyses in this book emphasize social demography, which uses population data as an empirical base to examine both the social causes of specific demographic realities and the social consequences of those realities. The social demography perspective also includes a humanistic orientation that points up practical problems in dealing with mortality, especially its causes. I have tried to bring these elements together into description, analysis, concern, and proposed action.

My intent was not to write a textbook, although I hope college and university professors will find the volume useful as a supplement for courses in social problems, demography, medical sociology, and other areas of the social sciences. The book is meant primarily to be a reference for university and other library collections to provide additional information for those readers already knowledgeable about mortality in the United States and to enable others to begin that study. Therefore, it contains extensive data compilations and explanations of what I think those data mean.

The book is organized into seven chapters. Chapter 1 provides several perspectives on mortality, especially the epidemiologic transition in the United States, and the biological and environmental components of mortality, reflected in life span and life expectancy. The chapter then discusses the measurement of mortality, including the quality of available data and several indexes used to compute and report mortality patterns and trends.

Chapter 2 deals briefly with general death rates but is largely a study of mortality differentials. That analysis includes differences by age and it uses age controls to make several mortality comparisons, especially between developed and developing countries and between the United States and other industrialized nations. Mortality differentials within the United States include those by sex, race and Hispanic origin, marital status, and socioeconomic status as measured by several social indicators. The chapter accounts for geographic variations in mortality using age-adjusted death rates by race and sex, and it ends with trends in differential mortality.

Chapter 3 explores variations in infant mortality. The analysis makes some international comparisons, contrasting infant mortality in developed and developing countries and comparing the United States with other developed countries. Then come patterns in the United States by age of the victim, including the perinatal, neonatal, and postneonatal periods. The discussion moves to variations in infant mortality by sex of the victim and deals briefly with maternal mortality, which is closely related to infant mortality. The chapter considers differentials by race and ethnicity—comparisons that enable inferences about differences in the average well-being of various groups. The chapter shows the geographic distribution of infant mortality by region, division, state, county, and urban and nonurban residence. The analysis includes major causes of death among fetuses and infants; it concludes with trends.

Chapter 4 focuses on cause of death, using the underlying cause as the basic datum and emphasizing the complex interaction among causes. It includes differentials and geographic distribution. The first part of the analysis identifies the ten leading causes of death in the United States and also includes a section on HIV/AIDS, drugs, alcohol, and Alzheimer's disease as factors in mortality patterns. The chapter treats variations in cause by sex, race and Hispanic origin, and age. It concludes with the geographic distribution of deaths from heart disease and malignant neoplasms (cancer).

Chapter 5 accounts for trends in cause of death and focuses first on reductions in the virulence of most infectious and parasitic diseases. It then deals with trends in cardiovascular diseases, including various forms of heart disease and several contributing conditions—high blood pressure, smoking, high serum cholesterol, and obesity. The section on cardiovascular diseases also includes trends in mortality for stroke. The chapter traces mortality trends for various forms of cancer and for diabetes mellitus, pneumonia, and several other diseases, as well as trends in environmental causes—accidents, suicide, and homicide. The chapter concludes by examining trends in male/female and black/white mortality ratios for selected causes.

Chapter 6 analyzes life expectancy. It elaborates the concept of life expectancy and examines differentials by age, sex, and race, together with the impact of widowhood on women because of their longer life expectancy. The chapter includes geographic variations in life expectancy and state-to-state differentials by race and sex. It accounts for the relationship between life expectancy and cause of death, especially the probable effects on longevity of eliminating certain causes. The chapter looks at potential years of life lost to certain causes of death, and at the concept of "active life expectancy," especially among the elderly. There is also a discussion of national health goals and life expectancy. The chapter continues with a section on long-term and short-term longevity trends and concludes with some long-range projections.

Chapter 7 summarizes the book's principal conclusion about mortality patterns and trends in the United States.

I am indebted to the many people who gather the data that make analyses such as this book possible. They include physicians and others who record the facts of each individual death, persons who compile the millions of individual reports into aggregate data, and those who cross-classify the data for the nation and its various subgroups and geographic divisions.

I also owe a great debt to the countless scholars who analyze the statistics on various aspects of mortality. They include professionals at the Centers for Disease Control, the National Institutes of Health, the National Center for Health Statistics, the U.S. Bureau of the Census, and many colleges, universities, and research institutes. Without their insightful and penetrating work, my project would be impossible. I hope I have credited them fairly and done justice to their painstaking, cumulative efforts.

As always, I deeply value the support and encouragement of my colleagues and students at Guilford College, and I am indebted to the college itself for a sabbatical leave and a research grant that helped make this project possible. I especially thank Kathrynn Adams, Vernie Davis, Chris Gjording, William Rogers, Sam Schuman, Carol Stoneburner, and the many students who listened to my accounts of the ongoing research and offered valuable insights and evaluations.

In this project as in others, my work is strongly influenced by the late T. Lynn Smith, whose rigorous methods, far-ranging social demography, and human concern continue to inspire what I do.

I offer my gratitude to the many persons at Greenwood Press who have contributed to the finished product. They are a very professional group with high standards of excellence, and I feel privileged to continue working with them.

I greatly appreciate the painstaking work of Bryan Boulier of George Washington University, who read the entire manuscript and offered many valuable suggestions.

As ever, I give my deepest thanks and devotion to Evelyn Zopf, whose love and patience make it possible for me to pursue my research interests and whose constant encouragement makes my writing worth doing, regardless of whether it is commended by others.

I thank Eric Zopf for his support of my work and for his interest in the results.

I hope this book lives up to the high expectations of all these people and that it will prove useful to many persons interested in the health and mortality of Americans.

1 Introduction: Perspectives on Mortality

This book is about an American revolution comprising lower general death rates, dramatically reduced infant mortality rates, increased life expectancy, changes in the relative importance of various causes of death, shifting mortality differentials among population components, and even the prospect for increases in the "fixed" life span. It is also about reversals, periodic setbacks, and inequitable participation by various groups in the downward mortality trend. Furthermore, the long-term American experience is part of a major worldwide phenomenon in which decreases in death rates, combined with stubbornly high birth rates, produced rapid population growth.[1] The mortality decreases accelerated in Western Europe and the United States during and after the latter part of the eighteenth century, although in some places they actually began earlier; they got seriously underway in the developing countries during the second half of the twentieth century. The population growth caused major changes in social institutions, patterns of interaction, resource use, and the equilibrium between population and environment, though at different rates and under different circumstances in various places. The effects of declining mortality rank with other major socioeconomic revolutions, such as changes in agriculture and industry; large-scale urbanization in predominantly rural populations; cybernation; and the advent of nuclear weapons. Indeed, not only are mortality reductions part of a much larger sociocultural restructuring, they are partly responsible for it. Many social changes would have been smaller and slower without the dramatic population increase made possible by falling death rates. Some would not have been significant at all, including burgeoning child dependency burdens, large influxes of young adults into job markets, and, in countries where rates of population growth have slowed or even stopped, aging of the population. That aging now typifies America's population because of decreases in mortality coupled with high birth rates when today's elderly were young and low birth rates now, thereby making children a smaller percentage of the population and elderly persons a larger share.

Thus, reductions in mortality are a phenomenon, a cause, and an effect. They are one of many complex demographic changes in the United States and elsewhere; they have produced changes in many social, cultural, economic, political, and other subsystems; they are the result of other changes, especially improvements in and more widespread knowledge about nutrition, sanitation, and preventive and curative medicine.[2]

EPIDEMIOLOGIC TRANSITION

Death controls that cause the shift from high to low mortality rates, though unevenly available to various countries and groups within countries, underlie the *epidemiologic transition*—a revolution in its own right. Described by Abdel R. Omran, the transition includes three stages:

1. "The Age of Pestilence and Famine" has high and wildly fluctuating mortality rates, frequent pandemics, and life expectancy between 20 and 30 years. It was humankind's common condition from prehistory until the eighteenth century; it prevailed in the developing countries until the second half of the twentieth century and persists in some.

2. "The Age of Receding Pandemics" begins when mortality peaks are lower and less frequent and general mortality begins to fall. Movement through this stage accelerates as the great epidemics become less frequent and less virulent and as mortality, especially among infants, falls fairly rapidly. Life expectancy at birth increases to 55 years or so. This stage is marked by rapid population growth because birth rates rarely decline at the same pace as death rates, and the gap between the two usually widens significantly. Many developing countries, especially in sub-Saharan Africa, are in this stage now and others were in it quite recently.

3. "The Age of Degenerative and Man-made Diseases" reflects low and relatively stable mortality. The principal causes of death (e.g., heart disease, cancer, stroke) typify a population with a rising median age and a growing percentage of elderly people. In addition, the society usually has an industrial system in which accidents, environmental pollution, and other dysfunctional by-products of sophisticated technology take a significant toll of lives. Contagious illnesses are well controlled most of the time, some almost to the vanishing point, and life expectancy is 70 years or more. The birth rate is usually low for countries in this stage, but it may rise somewhat to produce modest growth at times, only to fall and create zero growth or decline at others. All industrialized nations are in this stage, though under conditions that cause their respective birth and death rates to vary. Medical advances are especially important in saving lives, unlike the first two stages in which better nutrition and hygiene often provide more effective death control than does evolving medical technology.[3]

Epidemiologic Transition in the United States

Progress through the epidemiologic transition in the United States during the twentieth century is described under trends in most chapters that follow. Those trends, however, can be traced only from 1900 to the present. Despite that limitation, the data reflect a nation that finished two final decades in the age of receding pandemics, exemplified by the influenza epidemic of 1917–18, and then moved into the age of degenerative and human-made diseases with a relatively stable but still gradually declining mortality rate. While it is difficult to document the situation prior to 1900, fragmentary data for Massachusetts after 1800 and for New York City after 1804 suggest that "mortality in the United States in the eighteenth century and until the mid-nineteenth century was generally high with fluctuations in epidemic years [though] evidence indicates that the mortality decline in the United States as a whole occurred earlier than in New York City."[4] Thus, in the late 1700s the nation was still in but moving toward the end of the first stage of the epidemiologic transition. Even so, as recently as 1900 pneumonia, influenza, tuberculosis, diarrhea, and enteritis— all typical of inadequate health protection—caused 34 percent of the nation's deaths; heart disease and cancer combined caused only 12 percent. In 1988, well into the last stage of the transition, the first group of illnesses caused less than 5 percent of all deaths, while heart disease and cancer caused more than 58 percent.[5]

Succeeding chapters explore various aspects of mortality in the last stage of the epidemiologic transition, though trends also show how the United States moved out of the final part of stage two and well into stage three. That movement appears as trends in (1) age-specific and age-adjusted death rates; (2) mortality differentials by sex, race, and other characteristics; (3) infant mortality; (4) the relative importance of various causes of death; and (5) life expectancy at birth and other ages.

BIOLOGICAL AND ENVIRONMENTAL MORTALITY COMPONENTS

Biological influences on mortality fall under two fundamental concepts. They are *life span*, which is an expression of how long members of the human species can possibly survive, and *life expectancy*, or *longevity*, which describes how long persons now alive can avoid various hazards, resist death, and remain alive. Life span supposedly is an immutable characteristic governed by the species' genetic make-up, whereas life expectancy is affected not only by biological determinants, but by many environmental influences as well.[6]

Life Span

If the environment, including prenatal conditions, is completely favorable, "the time of a person's death is determined by genetic endowment and death is due to the influence of biological forces."[7] In turn, that genetic endowment is determined by chance and by the genetic make-up of one's biological antecedents, so it varies from person to person. This "biological clock of aging" is also affected by the apparent inability of normal cells to reproduce more than a certain number of times, probably 50 or 60 or so—the so-called "Hayflick effect"— and by other complex processes and limitations built into all species. Therefore, despite individual variations, presumably there is an upper limit on the survival potential of even the healthiest, most environmentally safe and genetically durable humans.[8]

We really do not know the upper limit of the life span, but it should be about the length of life of the oldest person accurately recorded. By 1991, that person was Shigechiyo Izumi of Asan, Tokunoshima Island, Japan, who was documented to have lived 120 years, 237 days, before he died in 1986. One other person was accurately reported to have lived to 114, one to 113, five to 112, six to 111, and seven to 110.[9] There are reports of persons who lived much longer than 120 years, but none has been proven accurate and many have turned out to be deliberate exaggerations or inadvertent errors. Thus, the present absolute limit of the life span appears to be no more than 120 years; 115 years may be more realistic.[10]

Some researchers doubt that life span is fixed and that the absolute limit actually is 115–120 years. Efforts are underway to intervene in aging, including genetic engineering to alter processes that cause cells, organs, the whole bodily systems to age, along with efforts to regenerate tissue so as to modify the Hayflick effect. These are *life-span-extending technologies*, and while they are much more difficult to achieve than disease prevention and control, they could push the life span upward to some new limit, perhaps 125 years or more. The dashed line in the upper part of Figure 1–1 shows the possible results of implementing such technologies. The solid line shows present life expectancy at various ages.

If new technologies are to extend the life span, they will have to be applied early in life so as to counteract genetic programming for decline and death. The technologies would mollify long-term effects of poor diet, temperature extremes, progressive breakdown of the immune system, loss of tissue elasticity, and finite cell replication. They would also slow accumulation of harmful cell substances, abnormal oxidation, membrane damage, and related conditions that influence aging from the first moment of life.[11] Life span cannot become infinite, of course, but the technology to extend it, especially by manipulating the immune system, now exists or soon will. That could be one result of research on acquired immune deficiency syndrome (AIDS). Therefore, the assumption that the human life span is biologically fixed at 115–120 years may prove outdated, and advances in the twenty-first century may lower the death rate and increase the numbers of very elderly people by altering an apparently inalterable species attribute. Numbers of persons 85 years old and over are already increasing faster than any other age group, partly because of improvements in health and partly because of fertility levels that prevailed when they were born; any new forces that expand their age

Figure 1–1
Effect of Life Span-Extending and Curve-Squaring Technologies

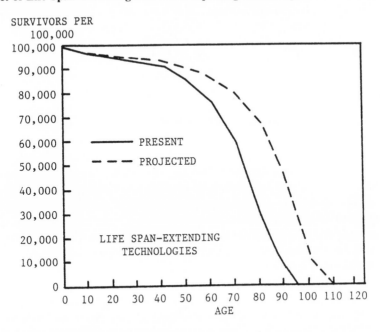

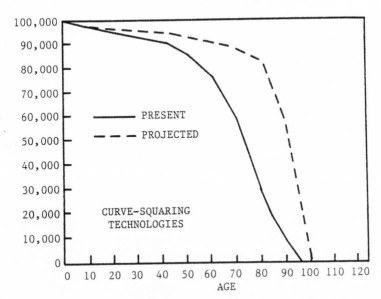

Source: U.S. House of Representatives, Subcommittee on Human Services of the Select Committee on Aging, *Future Directions for Aging Policy: A Human Services Model* (Washington, DC: Government Printing Office, 1980), p. 110.

group even faster will have profound social, economic, and other consequences. A major issue then will be not whether the average life can be lengthened, but what quality that longer life will have.

Life Expectancy

No matter what the prospects for extending the life span, increases in life expectancy already are impressive, with significant demographic and social effects. In 1900, for example, average life expectancy at birth for both sexes combined was 47.3 years, whereas in 1988 it was 74.9 years. In 1900 only 4.1 percent of the population was 65 and over, compared with 12.3 percent in 1988, and while the total population slightly more than tripled during the period, the elderly group grew almost tenfold. This was the result not only of reductions in mortality, of course, but also of fluctuations in fertility, which caused the relative size of various age cohorts eventually entering the older ages to vary considerably. Changes in the numbers of immigrants from year to year had a similar effect, swelling the elderly population more in some periods than others. Despite the increases in life expectancy, there is still room for more because the 1988 average was less than two-thirds the assumed life span. Thus, actual longevity is well below potential longevity because of individual genetic make-up, susceptibility to various diseases, exposure to hazards, and other biological and environmental conditions that winnow out persons of all ages, from embryos, fetuses, and infants, to centenarians.

Environmental conditions that affect the incidence of infectious and parasitic illnesses, the availability of medical treatment, and the degree of exposure to accidents, homicide, and other violences are at least as important as genetic composition in creating longevity differentials. In most societies females at all ages have greater life expectancy than males, but there are exceptions and the longevity gap between the sexes varies widely from society to society, often because of the status of females and the ways they are treated. Therefore, while the biology of sex plays an important part in differential longevity, environmental conditions also are significant. Moreover, the genetic composition of various racial, ethnic, and other groups makes some more susceptible than others to certain causes of death, such as sickle-cell anemia, which is almost exclusive to black people and which often kills at a young age. Such genetic influences on mortality are small, however, compared with the impact of a higher poverty rate among blacks, their poorer average medical treatment, and their incidence of low-weight and immature births, especially to very young mothers. Those and other conditions in the sociocultural environment account for much more of the racial differential in life expectancy than do genetic variations between the two groups.

Whatever one's genetic potential for longevity, he/she can modify its effects. Minimizing stress seems to increase longevity, as do optimistic attitudes about life, a good diet, regular exercise, and normal or slightly lower weight. Also important are avoiding smoking, using alcohol only moderately, sleeping seven or eight hours, and eating regular meals, particularly breakfast. In fact, throughout the world many groups that report unusually long average life expectancy actually have lifestyles that increase life expectancy even while life span remains within the species limit. Many well-publicized cases of extremely long life, such as the Abkhasians in Russia, are due to healthful lifestyles, even after accounting for data inaccuracies.[12]

Table 1-1 suggests the impact on longevity of a nation's sociocultural environment, especially its ability to implement death controls for newborns. It shows average life expectancy at birth for the world as a whole, for more developed and less developed countries, and for one group of nine countries where life expectancy is 77 years or more and another nine where it is 45 years or less. The United States is included for comparison. The table also provides some social indicators (per capita Gross National Product, infant mortality rate, and percentage of people classified as urban). These indicators are not meant to show that specific increases in GNP produce particular incremental improvements in life expectancy or that increases in the percentage of urban dwellers are necessarily associated with extended longevity in every country. Moreover, the three indicators are only a few of many that could be used to measure socioeconomic well-being. The precise influence of each also varies within its own cultural context, which, in turn, represents a highly complex interplay of forces. The data do show, however, that socioeconomic conditions have an important impact on life expectancy and that they greatly modify genetic prospects for survival.

To the extent death controls improve worldwide, life expectancy should increase significantly, especially where it is still quite low. In 1988, for example, 64 countries with 1 million or more persons had life expectancies below the world average of 63 years; 51 nations ranked above average. As countries in each stage of the epidemiologic transition improve death controls, we can anticipate lower mortality and higher life expectancy, though perhaps with temporary reversals. Even in Japan, where life expectancy is the world's highest (78 years), it can still increase. In countries such as Sierra Leone and Ethiopia, where the average newborn can expect to live only 41 years, life expectancy can increase significantly. If significant increases occur in life span and life expectancy, the world's population will continue to grow rapidly as long as birth rates remain high, although they are more likely to fall and populations to age. In either case, mortality reductions and greater length of life will affect social institutions, governmental programs to meet the needs of various groups, economic and political conditions, and the nature and magnitude of the support burden that producing adults must carry.

In developed countries, applying additional lifesaving technology to middle-aged persons would increase the number who live to the older ages, especially the 80s and 90s, and would do so in a relatively short time.[13] Such efforts would use *curve-squaring technologies*, which include prevention, diagnosis, and treatment of the three major causes of death; use of artificial organ substitutes;

Table 1-1

Life Expectancy at Birth and Scores on Three Social Indicators, Selected Countries, 1990

Category of Country	Life Expectancy at Birth (Years)	Per Capita GNP ($US)[a]	Infant Mortality Rate[b]	Percent Urban
World	64	3,470	73.0	41
More developed	74	15,830	16.0	73
Less developed	61	710	81.0	32
Life expectancy of 77 years or more				
Japan	79	21,040	4.8	77
Canada	77	16,760	7.3	77
France	77	16,080	7.5	73
Greece	77	4,790	11.0	58
Hong Kong	77	9,230	7.4	93
Netherlands	77	14,530	7.6	89
Spain	77	7,740	9.0	91
Sweden	77	19,150	5.8	83
Switzerland	77	27,260	6.8	61
Life expectancy of 45 years or less				
Angola	45	500[c]	137.0	25
Guinea-Bissau	45	160	132.0	27
Liberia	45	450	83.0	43
Niger	45	310	135.0	16
Somalia	45	170	132.0	33
Guinea	42	350	147.0	22
Afghanistan	41	250[c]	182.0	18
Ethiopia	41	120	154.0	11
Sierra Leone	41	240	154.0	28
United States	75	19,780	9.7	74

Source: Population Reference Bureau, 1990 World Population Data Sheet (Washington, DC: Population Reference Bureau, 1990); John Paxton, ed., The Statesman's Yearbook, 1990-91 (New York: St. Martin's Press, 1990), pp. 63, 81.

[a]For 1988.

[b]Deaths of infants aged 0-1 per 1,000 live births.

[c]For 1985.

coronary-artery surgery, angioplasty, and other techniques; and improvements in the environment, such as stress reduction, dietary modification, and further decreases in the proportion of smokers. These technologies reduce degenerative diseases and extend the life expectancy of older people. This contrasts with efforts earlier in the epidemiologic transition to curtail diseases caused by infections, parasites, and malnutrition, thereby increasing the percentage of new-borns who survive to the older ages. The lower part of Figure 1–1 shows how the curve-squaring technologies might change the number of survivors among each 100,000 Americans in specific age groups. With better controls for heart disease, cancer, stroke, and other degenerative diseases, life expectancy might rise to 85 by 2025, and the number of elderly people would be larger than now projected; so would their percentage if no new baby boom occurred.[14]

Many specific aspects of the curve-squaring technologies are already well advanced and will probably improve significantly in the early 2000s.[15] The present virulence of the three major causes of death may even be reduced more substantially within 50 years, thus adding to the elderly population. Regardless of the specifics, increases in life expectancy among adults are continuing and helping the very elderly to increase faster than any other age group.[16]

MEASURING MORTALITY

This section presents indexes used throughout the book to manipulate data on mortality. The data themselves come largely from reports of the National Center for Health Statistics (NCHS), which is part of the Public Health Service operating within the U.S. Department of Health and Human Services. In turn, the data are derived from the national system for the registration of vital statistics. Methods used to collect and report the data and indexes used to measure mortality have strengths and weaknesses, and while other sections of the chapter discuss specific problems, here we consider briefly some overall difficulties.

Quality of the Data

An early problem one encounters in using and interpreting figures on mortality is the indistinct line between two seemingly distinct states—life and death—because medical intervention has made it more difficult to ascertain for statistical purposes when life begins and ends. Therefore, to provide working concepts the vital statistics system distinguishes between live births, deaths, and fetal deaths.[17] These distinctions seem obvious and the events mutually exclusive, but in practice it is often difficult to separate them into distinct categories, as subsequent sections show.

Registration Area. Data on long-term trends may be somewhat misleading because of changes in the vital registration area. The first federally published death statistics were collected in the census of 1850, so even the most rudimentary national efforts to gather materials on mortality are less than a century and a half old. In 1880 a national "registration area" was created, but it consisted only of Massachusetts, New Jersey, the District of Columbia, and several large cities with fairly efficient data-gathering systems. The mortality data were reported in 1900 by ten states and the District of Columbia and in later years by a growing number of death-registration states, until 1933 when the registration area came to include the entire coterminous United States. It now includes all 50 states and the District of Columbia.[18] Thus, materials prior to 1933 refer to only part of the population, and while trends since 1900 are useful in several respects, they are derived from an incomplete data base.

Cause of Death. Errors and misinterpretations are most likely in reporting the cause of death, and they may occur in several ways. While a particular death becomes part of the aggregate data according to a single cause, many deaths result from several interacting causes. Consequently, the individual death certificate, from which aggregate mortality data are derived, allows the death certifier to specify not just the immediate cause of death, but also contributing causes, and it emphasizes what is diagnosed as the *underlying cause* (see Figure 1–2). For example, the immediate cause of death of a very elderly person may be pneumonia, but the underlying cause may be congestive heart failure, and a contributing cause of the heart problem may be atherosclerosis, or the long-term accumulation of deposits on arterial walls. Thus, the victim died of an illness contracted perhaps only a few days before death, partly because of weakness and susceptibility from the diminishing capacity of the heart to pump blood, which occurred largely because of a lifetime of clogging in the circulatory system.

The death certificate also allows other significant causes than the underlying one to be reported, such as accidents and injuries. Thus, the elderly person suffering from pneumonia and from congestive heart failure as the underlying cause of death may have tried to get out of bed, fallen, and died of accidental injury. In the individual case, the complex interplay of causes of death can be inferred from the death certificate, but in aggregate data for whole populations and major subcategories within them, we are generally left with a single cause of death—the underlying one that contributed most significantly to the victim's demise.[19]

"Since 1949, cause-of-death statistics have been based on the underlying cause of death, which is defined as '(a) the disease or injury which initiated the train of events leading directly to death, or (b) the circumstances of the accident or violence which produced the fatal injury.' "[20] This emphasis on the underlying cause ensures uniformity in reporting and analyzing the events that precipitate death. The causes are standardized by the World Health Organization (WHO) in its *Manual of the International Statistical Classification of Diseases, Injuries, and Causes of Death (ICD)*, which describes the specific causes and groups them into major categories and specific subcategories at several levels.[21] The *ICD* also includes rules for systematically identifying the underlying cause of death, which is relatively easy to report and understand and which provides some insight into the chain of events that resulted in death. To that extent the *ICD* does provide uniformity and comparability. On the other hand, the *ICD* has been revised several times, and while the revisions introduce refinements and categories that correspond better to changing realities, they create certain comparability problems over time. In addition, revisions go on even under an existing version. The Ninth Revision, first applied to data for 1979, includes several coding changes

Figure 1-2
Standard Certificate of Death, 1989 Revision

NORTH CAROLINA DEPARTMENT OF ENVIRONMENT, HEALTH, AND NATURAL RESOURCES
DIVISION OF EPIDEMIOLOGY — VITAL RECORDS SECTION

CERTIFICATE OF DEATH

Registration
District No. _____ Local No. _____

DECEDENT

1. DECEDENT'S NAME (First, Middle, Last)

2. SEX

3. DATE OF DEATH (Month, Day, Year)

4. SOCIAL SECURITY NUMBER

5. AGE — Last Birthday (Years)

5b. UNDER 1 YEAR — Months / Days

5c. UNDER 1 DAY — Hours / Minutes

6. DATE OF BIRTH (Month, Day, Year)

7. BIRTHPLACE (City and State or Foreign Country)

8. WAS DECEDENT EVER IN U.S. ARMED FORCES (Yes or No)

9a. PLACE OF DEATH (Check only one: see instructions on other side)
HOSPITAL: □ Inpatient □ ER/Outpatient □ DOA OTHER: □ Nursing Home □ Residence □ Other (Specify)

9b. FACILITY NAME (If not institution, give street and number)

9c. CITY, TOWN, OR LOCATION OF DEATH

9d. INSIDE CITY LIMITS? (Yes or No)

9e. COUNTY OF DEATH

10. MARITAL STATUS—Married, Never Married, Widowed, Divorced (Specify)

11. SURVIVING SPOUSE (If wife, give maiden name)

12a. DECEDENT'S USUAL OCCUPATION (Give kind of work done during most of working life. Do not use retired.)

12b. KIND OF BUSINESS/INDUSTRY

13a. RESIDENCE — STATE

13b. COUNTY

13c. CITY, TOWN, OR LOCATION

13d. STREET AND NUMBER

13e. INSIDE CITY LIMITS (Yes or No)

13f. ZIP CODE

14. Was Decedent of Hispanic Origin? (Specify Yes or No—If yes, specify Cuban, Mexican, Puerto Rican, etc.) □ Yes □ No (Specify)

15. RACE — American Indian, Black, White, etc. (Specify)

16. DECEDENT'S EDUCATION (Specify only highest grade completed) Elementary/Secondary (0-12) College (13-17+)

PARENTS

17. FATHER'S NAME (First, Middle, Last)

18. MOTHER'S NAME (First, Middle, Maiden Surname)

INFORMANT

19a. INFORMANT'S NAME (Type/Print)

19b. MAILING ADDRESS (Street and Number or Rural Route Number, City or Town, State, Zip Code)

Figure 1-2 (continued)

CAUSE OF DEATH

PART I. Enter the diseases, injuries, or complications that caused the death. Do not enter the mode of dying, such as cardiac or respiratory arrest, shock or heart failure. List only one cause on each line.

		Approximate Interval Between Onset and Death
IMMEDIATE CAUSE (Final disease or condition resulting in death)	a. DUE TO (OR AS A CONSEQUENCE OF):	
Sequentially list conditions if any, leading to immediate cause. Enter **UNDERLYING CAUSE** (Disease or injury that initiated events resulting in death) **LAST.**	b. DUE TO (OR AS A CONSEQUENCE OF):	
	c. DUE TO (OR AS A CONSEQUENCE OF):	
20a.	d.	

PART II. Other significant conditions contributing to death but not resulting in the underlying cause given in Part I.

20b.

AUTOPSY? (Yes or No)	If yes, were findings considered in determining cause of death?	Was case referred to Medical Examiner? (Yes or No)	TIME OF DEATH
21a.	21b.	21c.	22. M.

NOTICE: STATE LAW REQUIRES THAT ALL DEATHS DUE TO TRAUMA, ACCIDENT, HOMICIDE, SUICIDE, OR UNDER SUSPICIOUS, UNUSUAL, OR UNNATURAL CIRCUMSTANCES BE REPORTED TO, AND CERTIFIED BY A MEDICAL EXAMINER ON A MEDICAL EXAMINER'S CERTIFICATE OF DEATH. ANY DEATHS FALLING INTO THESE CATEGORIES IS WITHIN THE MEDICAL EXAMINER'S JURISDICTION REGARDLESS OF THE LENGTH OF SURVIVAL FOLLOWING THE UNDERLYING INJURY.

CERTIFIER

SIGNATURE AND TITLE OF CERTIFIER | DATE SIGNED (Month, Day, Year)
23a. | 23b.

NAME AND ADDRESS OF PERSON WHO COMPLETED CAUSE OF DEATH (ITEM 20) (Type or Print)
24.

DISPOSITION

METHOD OF DISPOSITION ☐ Burial ☐ Cremation ☐ Removal ☐ Donation ☐ Other 25a.	PLACE OF DISPOSITION (Name of cemetery, crematory, or other place) 25b.		LOCATION — City or Town, State, Zip Code 25c.

NAME AND ADDRESS OF FUNERAL HOME	SIGNATURE OF FUNERAL DIRECTOR	LICENSE NUMBER
26a.	26b.	26c.
	SIGNATURE OF EMBALMER	LICENSE NUMBER
	26d.	26e.

REGISTRAR'S SIGNATURE	DATE FILED (Month, Day, Year)
27.	28.

DEHNR 1872
(Revised 10/89)
VITAL RECORDS

Source: North Carolina Department of Environment, Health, and Natural Resources (Raleigh, NC).

made since then: in 1983, for example, changes were made for acquired immune deficiency syndrome (AIDS) and human immunodeficiency virus (HIV) infections, and the changes were applied to the data from 1981 on. HIV/AIDS was first reported by the NCHS as a separate cause of death in the 1987 edition of *Vital Statistics of the United States*. Changes in coding poliomyelitis were also introduced in 1981. In addition, for 1982 the definition of "child" was changed, and it affects the way in which child deaths are attributed to child battering and other maltreatment. In short, while the *ICD* does systematize the data, periodic changes also create certain methodological problems, especially in comparing data from one major revision to another and in some cases even from year to year.[22] Nevertheless, this book adheres closely to the *ICD* in the analysis of 72 selected causes for general mortality and 61 causes for infant mortality, and it seeks as much comparability as possible in the categories and subcategories of underlying causes.

The underlying-cause-of-death approach was more acceptable when the United States was in earlier stages of the epidemiologic transition and the major causes were communicable epidemic diseases and infant mortality. Now that chronic and degenerative diseases of the third transition stage are the principal causes, the single-cause approach is complicated by many contributing conditions.[23] In earlier times death often did result from a single cause, but now it is more likely to be the final outcome of complex interactions among causes. Therefore, cause-of-death models ideally should account for (1) multiple conditions, (2) the likelihood that the cause of a person's death began relatively early in life and gradually worsened, and (3) the complex interaction of one's disease with each other and with the person's body during aging.[24] These ideals can be pursued by examining individual death certificates for samples of population groups, but aggregate data for the population as a whole and large subgroups within it still generally must use the concept of underlying cause.

Reporting and Coding. Inaccurate reporting of causes on the death certificate also creates problems. For example, while stroke is identified correctly as the cause of death in a large majority of stroke cases, the type of stroke (e.g., cerebral, hemorrhage, cerebral thrombosis) is less likely to be accurately reported. The resultant data enable acceptable analyses of patterns of stroke deaths among various groups, but less acceptable analyses of kinds of stroke.[25] The same problem applies to other causes of death, producing some errors in recording the deaths from large categories of disease (e.g., cancer, heart disease), but many more errors in reporting subcategories and contributing causes.

Qualifications of certifiers also vary, as do conditions under which they ascertain causes. For example, a complete autopsy may be performed in a hospital or medical examiner's laboratory, or a person's death may simply be recorded at home and attributed to a known, long-term illness that apparently finally claimed its victim. Family members may also misreport the age of the decedent, so that too many or too few deaths are attributed to given age groups. This problem is especially serious in the ages 85 and over and at its worst among centenarians.[26] For these and other reasons, the aggregate data can be no more accurate than the individual datum—the death certificate—although certain statistical manipulations do help compensate for known types of errors.

Since 1968 the United States has used a computer system to ascertain the underlying cause of death. The Automated Classification of Medical Entities (ACME) codes all items on the death certificate; they are then matched against decision tables to select the underlying cause of death and to minimize errors of interpretation.[27] This procedure makes for greater accuracy but still selects one underlying cause of death from multiple causes listed on the certificate, and the multiple causes do not appear in the aggregate data.

Reports of deaths of fetuses (stillbirths) and newborns present unique problems. Most states, for example, report fetal deaths only after a gestation of 20 weeks or more, that is, 20 weeks since the first day of the last normal menstrual period and the date the stillbirth occurs. Several states report fetal deaths for all periods of gestation, and still others report either after 20 weeks or when the fetus has attained a certain weight, which is measured in grams in some places and pounds and ounces in others. In addition, the precise age of the fetus may be difficult to determine and some stillborn fetuses may be disposed of without being recorded. Therefore, even fetal deaths that are supposed to be reported are actually undercounted, although the data are relatively complete for fetal deaths of 28 weeks' gestation or more.[28] Deaths of newborns also contain errors in coding racial, ethnic, and other categories. A Texas study, for example, showed that deaths in the first 28 days were underreported in families with Spanish surnames, and the error produced an artificially low mortality rate for newborns in that group.[29] Actually, we can assume some underreporting for all types of deaths prior to birth and at various ages in the first year, and also that there is little counterbalancing overreporting. The problem is greatest among the poorest populations with the least access to regular prenatal care, hospital delivery, and other protections that would either save the new life or at least ensure that its loss is recorded.

Other reporting and coding errors are possible, especially at four junctures from the individual death to manipulation of aggregate data: (1) when the certifier ascertains cause of death, (2) when the death certificate data are transcribed onto the official death notification, (3) when the underlying cause of death is determined from reported multiple causes, and (4) when the statistics are processed, published, and interpreted. All such errors eventually appear in the materials with which analysts must work, and even the best vital registration systems do not produce completely accurate statistics. Consequently, no analyses, including the ones in this book, can be better than their data base.

Place of Residence/Place of Occurrence. Most deaths are recorded by the victim's place of residence rather than by where the death took place, though some data on place of occurrence are reported as well. The focus on place of residence is to keep from artificially inflating death rates in places with hospitals, nursing homes, and other facilities where deaths often occur, and unrealistically deflating rates in small towns, villages, inner-city poverty areas, and other places that lack such facilities. If deaths were attributed only to place of occurrence, one could find such misleading situations as a high rate in a core urban county with a large medical center and generally good health care, coupled with a low rate in a surrounding ring of poor rural counties lacking those amenities. That pattern could mean only that many persons from the rural counties died at the medical center. Therefore, all mortality figures in this book are for place of residence, although the place of residence and occurrence are usually the same, as with persons who die at home or at hospitals and other facilities in counties where they reside.

Age Controls. Age is the most crucial variable in mortality analysis and no index that fails to account for it is very useful, especially in comparing populations with markedly different age profiles.[30] Unfortunately, at the international level the index most used to report mortality—the crude death rate—does not control for age, and one could erroneously conclude from its use that well-being is far better or worse in some countries than it is. This is illustrated by comparing Asia (excluding China) with Europe. The crude death rate in both is 10 deaths per 1,000 population, but life expectancy at birth is 60 years in Asia and 74 years in Europe, and the infant mortality rate is 88 in Asia but only 12 in Europe. Different age structures of the two regions, not uniform health conditions, account for the identical crude rates.

The need for age controls and several ways to incorporate them are highlighted in the following discussion of mortality indexes and examples from the United States.

Crude Death Rate

The crude death rate is simply a ratio between the number of deaths in one year and the enumerated or estimated total population in that year, and it gives only a general impression of mortality levels and changes.[31] Its formula, with a 1988 illustration, is:

$$\frac{\text{Deaths in a given year}}{\text{Total population at mid-year}} \times 1,000$$

$$\frac{2,167,999}{246,329,000} \times 1,000 = 8.8$$

Thus, the crude death rate in 1988 was 8.8 deaths per 1,000 existing population regardless of age, sex, race, ethnicity, or other characteristics, though crude rates could be computed according to any of those characteristics except age. When age differences are included, the death rate is no longer "crude," but a more refined way of measuring mortality.

Age-Sex-Specific Death Rate

The age-sex-specific death rate attributes deaths to the age ranges in which they actually occur, separately for the sexes. It can be computed for single years of age but is usually done for five- or ten-year age ranges. The formula, using a 1988 illustration for black women aged 25–34, is:

$$\frac{\text{Deaths of black women aged 25–34 in 1988}}{\text{Total number of black women aged 25–34 in mid-1988}} \times 100,000$$

$$\frac{4,518}{2,885,000} \times 100,000 = 156.6$$

Thus, there were 156.6 deaths of black women aged 25–34 per 100,000 in that race, sex, and age group (or 1.6 deaths per 1,000).

Age-specific death rates require relatively detailed data, and while the death certificate records them, there is considerable room for error. Age at death may be misreported among infants, for example, perhaps because the body is simply disposed of unofficially, is misidentified as a fetal death, or the loss is deliberately concealed if it jeopardizes the mother or her family. This is why deaths by child abuse are underreported. People also tend to "round off" ages, so that someone giving the age of an 84-year-old decedent may report 85; there is also a tendency to prefer the digits 0 and 5 and to avoid, 1, 3,and 9.[32] Degrees of reporting accuracy also vary among racial, ethnic, socioeconomic, and other groups. Yet, while age misreporting is a problem, it seems not to create gross distortions in the United States, except perhaps in the oldest ages, so the age-specific death rate gives a much more accurate impression of actual mortality among various groups than does the crude death rate. Even so, the age-specific rate is cumbersome to use: comparing several race and sex groups, for example, generates many figures because of age gradations.

Age-Adjusted Death Rate

The age-adjusted death rate distills age-specific death rates into a single figure, though a hypothetical one in certain respects. The age-adjusted rate indicates what the death rate would be in an actual population if it had the age distribution of a standard population. Thus, it is also a standardized death rate, computed by the *direct method of standardization*.[33] Any census year could be chosen for the standard population, but the NCHS uses 1940 to compute age-adjusted rates for the United States. The age-specific rates of an existing population in any other year are then applied to that standard population to show what the death rates would be if the existing population had the age distribution of the 1940 population. The procedure can be used to compare any populations or subcategories for which age-specific death rates are available. Table 1–2, for ex-

Table 1-2
Data to Compute Age-Adjusted Death Rates for Blacks and Whites, 1987

Age	Age-Specific Death Rate[a]		Standard Population 1940	Expected Number of Deaths	
	White	Black		White	Black
Under 1	845.1	2,003.7	2,020,174	17,072	40,478
1-4	46.4	82.1	8,521,350	3,954	6,996
5-14	24.1	33.9	22,430,557	5,406	7,604
15-24	93.8	135.0	23,921,358	22,438	32,294
25-34	115.7	263.1	21,339,026	24,689	56,143
35-44	184.2	480.4	18,333,220	33,770	88,073
45-54	451.9	923.0	15,512,071	70,099	143,176
55-64	1,182.1	1,907.1	10,623,336	125,578	202,598
65-74	2,688.9	3,668.6	6,325,058	170,074	232,041
75-84	6,247.8	7,313.9	2,278,373	142,348	166,638
85 and over	15,580.5	13,222.6	364,752	56,830	48,230
All ages	895.5	856.9	131,669,275	672,258	1,024,271

Source: U.S. Bureau of the Census, Sixteenth Census of the United States: 1940, Characteristics of the Population, United States Summary (Washington, DC: Government Printing Office, 1943), Table 7; National Center for Health Statistics, Vital Statistics of the United States, 1987, Vol. II, Mortality, Part A, (Washington, DC: Government Printing Office, 1990), Sec. 1, Table 10.

[a]Deaths per 100,000 population in specified group.

ample, provides the information to compute age-adjusted death rates for whites and blacks in 1987.

Columns 2 and 3 in Table 1-2 show the 1987 age-specific death rates for whites and blacks, respectively, and column 4 shows the 1940 standard population. To obtain the expected number of deaths among whites, their 1987 death rate at each age is multiplied by the 1940 standard population at that age, and the same is done for blacks, using their 1987 age-specific death rates and the 1940 population. The expected deaths then are summed separately for each race and the sum is divided by the total standard population and multiplied by 1,000 (or 100,000), which gives the age-adjusted death rate.[34] For whites in 1987 the age-adjusted death rate is:

$$\frac{672,258}{131,669,275} \times 1,000 = 5.1$$

For blacks it is:

$$\frac{1,024,271}{131,669,275} \times 1,000 = 7.8$$

Table 1-3
Mortality Measured by Three Indexes, by Race and Sex, 1987

Index	Total Population	All Races		White		Black	
		Male	Female	Male	Female	Male	Female
Crude death rate[a]	8.7	9.3	8.1	9.5	8.5	9.9	7.4
Age-specific death rate[b]							
All ages	8.7	9.3	8.1	9.5	8.5	9.9	7.4
Under 1	10.2	11.3	9.0	9.4	7.4	22.1	17.9
1-4	0.5	0.6	0.5	0.5	0.4	0.9	0.7
5-14	0.3	0.3	0.2	0.3	0.2	0.4	0.3
15-24	1.0	1.5	0.5	1.4	0.5	2.0	0.7
25-34	1.3	1.9	0.7	1.7	0.6	3.9	1.5
35-44	2.1	2.9	1.4	2.5	1.2	7.0	3.0
45-54	5.0	6.4	3.6	5.8	3.3	12.6	6.5
55-64	12.4	16.2	9.0	15.5	8.5	24.6	14.5
65-74	27.5	36.2	20.6	35.5	20.0	47.4	28.7
75-84	62.8	82.2	51.2	82.1	50.8	92.4	61.5
85 and over	153.2	180.3	142.6	184.3	144.9	152.3	123.1
Age-adjusted death rate[c]	5.4	7.0	4.0	6.7	3.8	10.2	5.9

Source: National Center for Health Statistics, Vital Statistics of the United States, 1987, Vol. II, Mortality, Part A (Washington, DC: Government Printing Office, 1990), Sec. 1, Tables 2-4.

[a]Deaths per 1,000 population of all ages.

[b]Deaths per 1,000 population in each age group.

[c]Deaths per 1,000 population computed by the direct method, using as the standard population the age distribution of the total population enumerated in 1940.

Thus, while the crude death rates of whites and blacks in 1987 were 9.0 and 8.6, respectively, the age-adjusted rates show a significantly different relationship because of greater socioeconomic problems, poorer average health care, and other conditions among blacks that elevate their mortality.

The importance of age controls is further underscored by the data in Table 1-3, which shows crude death rates, age-specific death rates, and age-adjusted death rates by race and sex. Crude death rates for the several groups are roughly similar, but when we account for variable age profiles of males and females because of differential mortality and for variable age profiles of blacks and whites because of differential mortality and fertility, the mortality indexes are strikingly dissimilar.

The nation's overall crude death rate is about 8.8 deaths per 1,000 population.

But age-specific death rates show how heavily mortality is concentrated in the older years, so when we account for the growing number and proportion of elderly people and their statistical influence on overall mortality, the age-adjusted rate drops to 5.4. Furthermore, the crude death rate for males is about 15 percent higher than for females, but since death rates are higher for males than females at all ages, the age-adjusted rate for males is 75 percent higher than for females. Similar variations exist by race: the crude death rates of white and black males are almost the same, but when the higher death rates of blacks at every age except 85 and over are taken into account, black males turn out to have an age-adjusted death rate 52 percent higher than white males. Black females have a crude death rate 13 percent below white females, but an age-adjusted death rate about 55 percent higher. This need to account for age variations in mortality is especially important in comparing a rapidly growing population that has a relatively youthful age structure with one that is growing slowly or not at all and which has a relatively high and increasing percentage of elderly.

Mortality Ratio

The mortality ratio compares death rates in one population with those in another, often by race, sex, and other characteristics. The ratio is expressed as one group's death rate, usually standardized for age, as a percentage of another group's rate. The formula, using the age-adjusted rates of black and white males in 1988, is:

$$\frac{\text{Age-adjusted death rate of black males}}{\text{Age-adjusted death rate of white males}} \times 100$$

$$\frac{1,037.8}{664.3} \times 100 = 156.2$$

Thus, the age-adjusted death rate among black males is 156.2 percent of that among white males.

Infant Mortality Rate

The infant mortality rate is commonly used to indicate the well-being of a population because any group that cannot look after the newborn adequately is likely to have low levels of living among a large share of its members and to neglect even rudimentary sanitation and nutrition. Infant mortality is particularly affected by the mother's health and the prenatal care she receives, and where such care is inadequate, infant mortality rates usually are high.

The infant mortality rate has a built-in age control, since it applies only to the first year of life. Moreover, refinements of the index pinpoint the most risky hours, days, and weeks during the first year; one variation even includes the last stages of pregnancy and the first six hazardous days of life. The infant mortality rate does not use the total population as a base, but the number of live births, so its accuracy depends on the care with which data on both births and deaths are gathered and reported.[35] With a 1988 illustration, the infant mortality rate is:

$$\frac{\text{Deaths of infants aged 0-1 in a given year}}{\text{Number of live births in that year}} \times 1,000$$

$$\frac{38,910}{3,909,510} \times 1,000 = 10.0$$

Thus, there were 10.0 deaths of infants for each 1,000 live births.

Perinatal Mortality Ratio. Deaths measured by the perinatal ratio occur just before, during, and just after birth, so the index separates a particularly vulnerable time from the preceding weeks of gestation and the postnatal infancy that follows. Because it includes part of the gestation period, the perinatal mortality ratio is not strictly a subdivision of infant mortality, but it does provide useful insights into the conditions of a high-risk time for mothers, fetuses, and newborns. Any duration of gestation and early infancy could be used for the index, and the NCHS uses three based on different combinations of gestation and days after birth. Definition I: infant deaths of less than seven days and fetal deaths with stated or presumed gestation of 28 weeks or more. Definition II: infant deaths of less than 28 days and deaths with stated or presumed gestation of 20 weeks or more. Definition III: infant deaths of less than seven days and fetal deaths with stated or presumed gestation of 20 weeks or more.[36] This book uses Perinatal Definition I, because the data on deaths after 28 weeks' gestation—*late fetal mortality*—are more accurate than those on deaths after 20 weeks.[37] The formula, with an illustration for 1987, is:

$$\frac{\text{Deaths under 7 days + late fetal mortality in a given year}}{\text{Number of live births in that year}} \times 1,000$$

$$\frac{38,164}{3,809,394} \times 1,000 = 10.0$$

Thus, in 1987 the perinatal mortality ratio was 10.0 deaths of infants under seven days and of fetuses with at least 28 weeks' gestation per 1,000 live births.

Neonatal Mortality Rate. The neonatal mortality rate refers to deaths from birth to 28 days of age, where almost two-thirds of all infant deaths are concentrated. Many deaths result from the same causes responsible for perinatal and late fetal mortality, but those causes often differ from the ones that produce most deaths later in the first year. The neonatal mortality rate and a 1988 example are:

$$\frac{\text{Deaths of infants under 28 days in a given year}}{\text{Number of live births in that year}} \times 1,000$$

$$\frac{24,690}{3,909,510} \times 1,000 = 6.3$$

Thus, there were 6.3 deaths of infants under 28 days of age for each 1,000 live births.

Postneonatal Mortality. Postneonatal mortality refers to the remaining deaths during infancy—from 28 days to the end of the first year. The risk of dying diminishes significantly and most losses that do occur result from different causes than the ones earlier in life (e.g., accidents versus low birth weight). The formula for the postneonatal mortality rate, with a 1988 example, is:

$$\frac{\text{Deaths of infants aged 4–52 weeks in a given year}}{\text{Number of live births in that year}} \times 1,000$$

$$\frac{14,220}{3,909,510} \times 1,000 = 3.6$$

Thus, 3.6 infants aged 4–52 weeks died per 1,000 live births, compared with 6.3 who died in just the first four weeks.

The infant mortality rate and its variations are approximations of reality, not true probabilities, because the deaths of infants (and late fetuses) refer to actual chronological age, whereas the number of live births refers to those during a calendar year. Therefore, some infants who died in 1988 were born in 1987. For example, a child born in mid-January 1987 and who lived the rest of that year and died in January 1988 before quite reaching its first birthday figures into the 1988 infant mortality rate. If the number of births remains about the same from year to year, this creates little aggregate distortion and conventional computations of infant mortality will reflect closely the probability of an infant dying in a given year. If there are significant fluctuations in births from one year to the next or even within a given year, however, the infant mortality rate and its variations might show a considerably better or worse mortality situation than actually exists.[38] Certain statistical methods compensate for this situation, but given the need to use official data that may lack such refinements, especially in international comparisons, these adjusted rates often are not available. Moreover, during the 1980s the number of births in the United States did not fluctuate widely from year to year, though it did rise, so the distortion is fairly minimal.

Maternal Mortality Rate

Deaths of women during childbirth or shortly thereafter as a result of complications are closely related to the deaths of infants, partly because both situations reflect the quality of care mothers and offspring receive during and after pregnancy and birth. More specifically, the Ninth Revision of the *ICD* identifies maternal mortality as the death of a woman while she is pregnant or within 42 days after the pregnancy is terminated by whatever means, if the cause is related to or aggravated by the pregnancy or the way it is managed. The concept does not include deaths from accidents or incidental causes. Any indirect causes, however, that might also be coded under other categories than maternal death, such as syphilis, tuberculosis, diabetes, drug dependence, and congenital cardiovascular disorders, are identified in the maternal category if they lead to the mother's death during pregnancy, childbirth, and the puerperium (the approximate six-week period from childbirth to the return of normal uterine size).[39] The maternal death rate, with a 1988 example, is computed on the basis of the number of live births:

$$\frac{\text{Deaths of women in a given year due to complications}}{\text{Number of live births in that year}} \times 100,000$$

$$\frac{330}{3,909,510} \times 100,000 = 8.4$$

Thus, for each 100,000 live births in 1988 there were 8.4 maternal deaths. The rate varies widely by age and is well below average for women under 30 and well above average for older women, reaching 73 for those 45 and over.

Cause-Specific Death Rate

The cause-specific death rate shows the number of persons per 100,000 population who die annually from each underlying cause and, therefore, the probability of dying from that cause. The formula is:

$$\frac{\text{Deaths from a given cause in a given year}}{\text{Total population at mid-year}} \times 100,000$$

For example, in 1988 there were 485,048 deaths from cancer and the population was estimated at 246,329,000.

$$\frac{485,048}{246,329,000} \times 100,000 = 196.7$$

Thus, 197 persons per 100,000 population died in 1988 from cancer. This is a crude rate, however, so age-specific and age-adjusted rates for each cause are used more commonly, often separately by race and sex.

Life Expectancy and the Life Table

Life expectancy is the average number of years of life remaining to persons at various ages, from birth through the oldest years. The *current life table* is used to compute the average (mean) number of person-years a group of people born in the same year or same five-year age span will live.[40] The group is an *age cohort* and the computed average years of life remaining at particular ages, derived from their mortality experience, is their life expectancy. A life table shows the life history of a hypothetical cohort of individuals exposed to a given set of age-specific death rates. The use of age-specific rates eliminates age fluctuations in actual populations and any changes in death rates they might incur because of disasters, major medical advances, or other reasons. Therefore, in constructing the life table it is assumed that mortality remains the same over the entire time anyone in the hypothetical cohort is still alive. In that sense the life table produces probabilities based on current mortality experience, and those probabilities, along with other information derived from the life table, make it one of the most useful tools in the study of mortality.[41]

Age-specific death rates, the number of survivors at a given age, and their remaining life expectancy could be presented by single years of age; such *complete life tables* are increasingly in demand for various purposes. The *abridged life table* is used more frequently, however, and it employs age categories 0–1, 1–5 . . . 85 and over. Overlap in the categories is avoided by assuming, for example, that the ages 1–5 comprise persons who have started the first day of the second year of life through those who have completed the last day of the fourth year but have not yet begun the first day of the fifth year. The abridged life table is illustrated by Table 1–4 for the total population of the United States in 1988. It should be noted that the reported deaths during 1988 are those actually recorded in that year, while the population used to compute the life table values is estimated as of July 1, 1988. This estimate is eight years removed from the 1980 census, and while various statistical techniques improve accuracy, the longer the time since an enumeration and the smaller the group being studied, the greater the margin for error. Therefore, data on life expectancy are subject not only to the same methodological problems as other indexes based on vital registrations, but also to difficulties with intercensal population estimates.

The values in Table 1–4 are interpreted as follows:

Column 1 (*x* to *x* + *n*) shows age intervals, or the period of life between two exact ages. For example, 15–20 refers to the five-year interval between the fifteenth and twentieth birthdays.

The basis of the life table is column 2 ($n^q x$) which converts age-specific death rates into proportions of persons alive at the beginning of an age interval

Table 1–4
Abridged Life Table for the Total Population, 1988

Age Interval	Proportion Dying	Of 100,000 Born Alive		Stationary Population		Average Remaining Lifetime
Period of Life Between 2 Exact Ages Stated in Years	Proportion of Persons Alive at Beginning of Age Interval Dying During Interval	Number Living at Beginning of Age Interval	Number Dying During Age Interval	In the Age Interval	In this and All Subsequent Age Intervals	Average Number of Years of Life Remaining at Beginning of Age Interval
(1)	(2)	(3)	(4)	(5)	(6)	(7)
x to $x + n$	nq_x	l_x	nd_x	nL_x	T_x	$\overset{\circ}{e}_x$
0–1	0.0100	100,000	999	99,147	7,494,642	74.9
1–5	0.0020	99,001	198	395,540	7,395,495	74.7
5–10	0.0012	98,803	120	493,688	6,999,955	70.8
10–15	0.0014	98,683	134	493,155	6,506,267	65.9
15–20	0.0044	98,549	431	491,767	6,013,112	61.0
20–25	0.0058	98,118	565	489,206	5,521,345	56.3
25–30	0.0061	97,553	596	486,274	5,032,139	51.6
30–35	0.0074	96,957	717	483,035	4,545,865	46.9
35–40	0.0096	96,240	924	479,021	4,062,830	42.2
40–45	0.0126	95,316	1,204	473,785	3,583,809	37.6
45–50	0.0189	94,112	1,777	466,443	3,110,024	33.0
50–55	0.0300	92,335	2,766	455,194	2,643,581	28.6
55–60	0.0473	89,569	4,238	437,859	2,188,387	24.4
60–65	0.0728	85,331	6,208	411,976	1,750,528	20.5
65–70	0.1055	79,123	8,344	375,656	1,338,552	16.9
70–75	0.1568	70,779	11,096	327,120	962,896	13.6
75–80	0.2288	59,683	13,654	265,113	635,776	10.7
80–85	0.3445	46,029	15,858	190,715	370,663	8.1
85 and over	1.0000	30,171	30,171	179,948	179,948	6.0

Source: National Center for Health Statistics, "Advance Report of Final Mortality Statistics, 1988," Monthly Vital Statistics Report 39:7 (Washington, DC: Government Printing Office, November 1990), Table 3.

(x) who will die before reaching the end of that interval ($x + n$). For instance, the proportion dying in the interval 15–20 is 0.0044, which means that of every 100,000 persons alive and exactly 15 years old, 44 will die before the twentieth birthday. Thus, the column expresses the probability that a person at the xth birthday will die before reaching the $x + n$th birthday.

Column 3 (l_x) shows the number of persons living at the beginning of the age interval (x) out of the total number of 100,000 births. In our example, out of 100,000 newborns, 98,549 survive to age 20.

Column 4 (n^{d_x}) shows the number of persons who die within the specified age interval (x to $x + n$) out of the 100,000 live births. In the example, there are 431 deaths in the ages 15–20 out of the initial cohort of 100,000 babies.

Column 5 (n^{L_x}) indicates the number of person-years lived during the age interval (x to $x + n$) by the cohort of 100,000. Thus, the 100,000 infants live 491,767 person-years between ages 15 and 20. Of the 98,549 persons who reach age 15, the 98,118 who survive to age 20 each live five years (98,118 × 5 = 490,590) and the 431 who die live varying periods averaging 2.5 years (431 × 2.5 = 1,078). The total of 491,668 person-years here is not exactly that in Table 1–4 (491,767) because the NCHS uses single years of age to compute the life table and then abridged it, so that total is slightly different from the one derived from the abridged life table.

Column 6 (T_x) shows the number of person-years lived after the beginning of the indicated age interval by the cohort of 100,000 births. In our example, the 100,000 babies live 6,013,112 person-years after their fifteenth birthday, that is, in the 15–20 and all subsequent age intervals. This figure is obtained by adding the ones in column 5, beginning with the oldest and proceeding to the youngest. Adding back from 85 and over through 15–20 produces 6,013,112 person-years.

Column 7 ($\overset{o}{e}_x$) is the average number of years of life remaining to a person who survives to the beginning of the age interval. Our example shows that someone who reaches the fifteenth birthday should expect an average remaining lifetime of 61.0 years. That average is obtained by the following formula:

$$\overset{o}{e}_x = \frac{T_x}{l_x}$$

Thus, for persons aged 15–20, the values are:

$$\overset{o}{e}_x = \frac{6,013,112}{98,549} = 61.0$$

This expectation of life at specified ages, derived from the life table, is commonly used to indicate the mortality situation in a population and, therefore, its general well-being.[42]

Computing life expectancy has rigorous data requirements, and while the necessary statistics are generally available for the nation and the states, subdivided by race and sex, at more local levels the requisite data often are not available. Even when they are, there may be so few deaths in a local area that they provide an inadequate data base for the painstaking task of constructing a life table.[43] Nevertheless, it would be useful to have life expectancy figures for counties, metropolitan statistical areas (MSAs), and other local populations, and to map these patterns to infer variations in health care and the level of living. Various methods yield such measures, but some assume stable death and birth rates and no migration in small areas. Others require more sophisticated data than are usually available. David Swanson, however, has developed a regression-based method to arrive at estimated life expectancy for small areas that requires only (1) total deaths in a county or other area, (2) estimated or enumerated population, and (3) estimated or enumerated population aged 65 and over. These data, available annually, yield the crude death rate and the percentage of persons 65 and over, and both are used to make local estimates of life expectancy.[44] The model needs more extensive testing to ascertain its usefulness for counties and other local populations, but it seems to have the accuracy needed for indexes of mortality ''to determine the health characteristics of small population groups'' and to locate places in which the incidence of cancer, cardiovascular diseases, and other causes of death is so high as to merit intensive epidemiological study and treatment.[45]

Mortality Differentials

Mortality differentials do not serve as an index in the same way that infant mortality rates, life expectancy, and others are used to measure the probability of death. Instead, differential mortality by age, sex, race, ethnicity, marital status, socioeconomic standing, and other characteristics depends on those indexes of probability. We can compare age-adjusted death rates for black males and white males, black females and white females, or any other combination of characteristics for which there are reliable data. We then can identify differentials among categories and draw inferences about comparative average socioeconomic standing, access to health care, and other conditions that affect mortality. Thus, differential mortality analyses have two purposes. First, despite major reductions in overall mortality in the United States, significant differences in death rates persist; some have actually increased. It is important to measure those variations. Second, since certain groups have not shared equitably in the improvements in longevity, it is imperative to identify some reasons why their progress has lagged and what might be done to lower their mortality further.[46] In fact, mortality levels in some U.S. groups are still very high compared with others and with the populations of other urban-industrial countries. Therefore, ''the goal of equal opportunity for all, so deeply ingrained in American ideology, tradition, and law, is still to be implemented in the realm of life itself—the achievement of equal opportunity for survival."[47] The study of differential mortality, using the indexes just described, is partly directed toward that end.

NOTES

1. A. H. Pollard, Farhat Yusuf, and G. N. Pollard, *Demographic Techniques*, 3d ed. (Sydney, Australia: Pergamon Press, 1990), p. 63.

2. Ester Boserup, *Population and Technological Change: A Study of Long-Term Trends* (Chicago: University of Chicago Press, 1981), pp. 124–125.

3. Abdel R. Omran, "Epidemiologic Transition: Theory," in *International Encyclopedia of Population*, Vol. 1, edited by John A. Ross (New York: Free Press, 1982), pp. 172–173. The original formulation is in Abdel R. Omran, "Epidemiologic Transition: A Theory of the Epidemiology of Population Change," *Milbank Memorial Fund Quarterly* 4:1 (October 1971), pp. 509–538. See also Regina McNamara, "Mortality Trends: Historical Trends," in Ross, *Encyclopedia*, Vol. 2, p. 461.

4. Abdel R. Omran, "Epidemiologic Transition: United States," in Ross, *Encyclopedia*, Vol. 1, pp. 175, 177.

5. National Center for Health Statistics, *Vital Statistics of the United States, 1987*, Vol. II, *Mortality*, Part A (Washington, DC: Government Printing Office, 1990), Sec. 1, Table 6; NCHS, "Advance Report of Final Mortality Statistics, 1988," *Monthly Vital Statistics Report* 39:7 (Washington, DC: Government Printing Office, November 1990), Table 10.

6. John R. Weeks, *Population: An Introduction to Concepts and Issues*, 4th ed. (Belmont, CA: Wadsworth, 1989), p. 150.

7. Bryan L. Boulier and Vincente B. Paqueo, "On the Theory and Measurement of the Determinants of Mortality," *Demography* 25:2 (May 1988), pp. 249–250.

8. The Hayflick effect was discovered by Leonard Hayflick, "Human Cells and Aging," *Scientific American* 218:3 (March 1968), pp. 32–37.

9. Donald McFarlan, ed., *The Guinness Book of Records, 1991* (New York: Facts on File, 1990), p. 13.

10. Edward L. Schneider and John D. Reed, Jr., "Life Extension," *New England Journal of Medicine* 312:18 (May 1, 1985), p. 1159. See also Josianne Duchene and Guillaume Wunsch, "From the Demographer's Cauldron: Single Decrement Life Tables and the Span of Life," *Genus* 44:3 (July-December, 1985), p. 5.

11. Theodore J. Gordon, "Prospects for Aging in America," in *Aging from Birth to Death*, edited by Matilda White Riley (Boulder, CO: Westview Press, 1979), p. 186.

12. Weeks, *Population*, p. 153.

13. U.S. House of Representatives, Subcommittee on Human Services of the Select Committee on Aging, *Future Directions for Aging Policy: A Human Services Model* (Washington, DC: Government Printing Office, 1980), pp. 109–110.

14. Ibid., p. 110. For projections of life expectancy to 2080, see U.S. Bureau of the Census, Gregory Spencer, "Projections of the Population of the United States, by Age, Sex, and Race: 1988 to 2080," *Current Population Reports*, Series P–25, No. 1018 (Washington, DC: Government Printing Office, 1989), p. 153.

15. Subcommittee on Human Services, *Future Directions for Aging Policy*, pp. 111–112.

16. For a study of this matter, see U.S. Congress, Office of Technology Assessment,

Life-Sustaining Technologies and the Elderly (Washington, DC: Government Printing Office, 1987).

17. Henry S. Shryock and Jacob S. Siegel, *The Methods and Materials of Demography*, Vol. 2 (Washington, DC: Government Printing Office, 1973), p. 389.

18. NCHS, *Mortality*, Sec. 7, p. 3.

19. Kenneth G. Manton and Eric Stallard, *Recent Trends in Mortality Analysis* (New York: Academic Press, 1984), p. 25.

20. NCHS, *Mortality*, Sec. 7, p. 8.

21. The version in use at this writing is the Ninth Revision (Geneva: World Health Organization, 1977).

22. NCHS, *Mortality*, Sec. 7, pp. 8–10.

23. Manton and Stallard, *Mortality Analysis*, p. 12.

24. Ibid. See also Robert R. Kohn, "Cause of Death in Very Old People," *Journal of the American Medical Association* 247:20 (May 28, 1982), pp. 2793–2797.

25. Manton and Stallard, *Mortality Analysis*, p. 20.

26. Ansley J. Coale and Ellen Eliason Kisker, "Mortality Crossovers: Reality or Bad Data?" *Population Studies* 40:3 (November 1986), p. 401.

27. Michael Alderson, *Mortality, Morbidity and Health Statistics* (New York: Stockton Press, 1988), p. 45.

28. NCHS, *Mortality*, Sec. 7, pp. 12–15.

29. Alderson, *Mortality, Morbidity*, p. 58.

30. Shryock and Siegel, *Methods and Materials*, p. 397. See also Pollard et al., *Demographic Techniques*, p. 66.

31. Shryock and Siegel, *Methods and Materials*, p. 397.

32. Ibid.

33. John Saunders, *Basic Demographic Measures* (Lanham, MD: University Press of America, 1988), pp. 46, 49. See also Pollard et al., *Demographic Techniques*, pp. 71–73.

34. The procedure is from NCHS, *Mortality*, Sec. 7, pp. 19–20; and Shryock and Siegel, *Methods and Materials*, pp. 418–421.

35. Pollard et al., *Demographic Techniques*, p. 68.

36. NCHS, *Mortality*, Sec. 4, p. 2.

37. Shryock and Siegel, *Methods and Materials*, p. 425.

38. Ibid., p. 411.

39. NCHS, *Mortality*, Sec. 7, p. 11.

40. Pollard et al., *Demographic Techniques*, pp. 30–31.

41. Ibid., p. 30.

42. The procedure for interpreting the life table is from Shryock and Siegel, *Methods and Materials*, pp. 431, 442. See also Saunders, *Basic Demographic Measures*, Chap. 6; Pollard et al., *Demographic Techniques*, Chap. 3; and NCHS, *Mortality*, Sec. 6, pp. 4–5.

43. Kenneth G. Manton and Eric Stallard, "Methods for the Analysis of Mortality Risks Across Heterogeneous Small Populations: Examination of Space-Time Gradients in Cancer Mortality in North Carolina Counties 1970–75," *Demography* 18:2 (May 1981), p. 217.

44. David A. Swanson, "A State-Based Regression Model for Estimating Substate Life Expectancy," *Demography* 26:1 (February 1989), pp. 161–162.

45. Manton and Stallard, "Mortality Risks," p. 217.
46. Evelyn M. Kitagawa and Philip M. Hauser, *Differential Mortality in the United States: A Study in Socioeconomic Epidemiology* (Cambridge, MA: Harvard University Press, 1973), p. 1.
47. Ibid.

2 General Mortality and Differentials

This chapter briefly discusses the general death rate, but it is largely a study of mortality differentials. They compare the United States with other urban-industrial countries and with some that are poorly developed. Within the United States the principal differentials are by age, sex, race and ethnicity, marital condition, and socioeconomic status. The chapter also shows geographic mortality variations for males and females in the United States, using the age-adjusted death rate.

AGE DIFFERENTIALS AND CONTROLS

Chapter 1 noted that mortality comparisons need to account for age variations. For example, the first year of life is relatively risky and the infant mortality rate and infant death rate are much higher than death rates among older children. Then, from its low level among children aged 5–14, the death rate climbs inexorably to its peak among persons 85 and over. This pattern appears in Table 2–1, which uses age-specific death rates by race and sex, averaged for a three-year period to minimize annual fluctuations.

Age profiles differ sharply among nations, especially those with recent population explosions and large percentages of infants, children, and young adults, and the ones with large numbers and percentages of elderly.[1] For instance, the crude death rates for males and females in the United States in 1985 were 9.5 and 8.1, respectively, while in Guatemala they were 9.4 and 8.0. Who would conclude from those similarities, however, that conditions which influence mortality are the same in the two countries? The crude death rates are similar because 46 percent of Guatemala's population is under 15—the age group with the lowest mortality rates even where levels of living are poor—whereas only 22 percent

Table 2–1

Age-Specific Death Rate, by Race and Sex, 1985–87

Age	White[a]		Black[a]	
	Male	Female	Male	Female
All ages	954.1	841.1	984.7	733.0
Under 1	985.9	763.0	2,176.0	1,759.7
1–4	52.2	40.3	90.1	73.6
5–9	26.2	18.2	39.2	29.6
10–14	33.7	19.2	44.7	23.7
15–19	118.1	48.3	133.6	47.4
20–24	159.6	50.2	247.4	79.7
25–29	155.1	54.3	316.3	115.5
30–34	174.4	67.2	439.9	175.5
35–39	211.2	95.4	589.6	238.0
40–44	290.9	151.7	741.9	355.7
45–49	452.2	252.1	986.9	517.7
50–54	751.2	418.0	1,452.5	800.0
55–59	1,229.8	656.7	1,995.1	1,169.5
60–64	1,944.6	1,052.5	2,959.1	1,804.3
65–69	2,910.9	1,602.4	3,950.6	2,401.0
70–74	4,594.6	2,518.6	5,627.3	3,530.7
75–79	6,996.9	3,999.3	7,702.0	5,008.3
80–84	10,832.0	6,810.9	11,729.5	8,309.9
85 and over	18,600.0	14,523.1	15,030.3	12,326.1

Source: National Center for Health Statistics, Vital Statistics of the
United States, 1985, 1986, 1987, Vol. II, Mortality, Part A (Washington,
DC: Government Printing Office, 1988–90), various tables.

[a]Deaths per 100,000 population in specified group.

of the United States population is under 15. Conversely, persons 65 and over, who have the highest death rates, are less than 3 percent of Guatemala's total but over 12 percent in the United States. If the 1985 United States population is used as a standard to compute age-adjusted death rates, they remain 9.5 for males and 8.1 for females in the United States, but in Guatemala they jump to 14.1 and 15.4, respectively, and give a much more accurate impression of actual mortality levels and the socioeconomic conditions that affect them.

SOME INTERNATIONAL COMPARISONS

The age factor will soon make crude death rates in most developed nations higher than those in most developing nations, reversing the historical relationship, even though death rates at each age are higher in the developing than the developed nations. Even in 1987 the crude death rate (10) in the developing countries collectively was the same as in the developed nations. The latter have higher life expectancy than ever, their fertility levels and percentages of children are relatively low, and their annual reductions in mortality are less than in the developing countries. These relationships between the two groups of countries will pass, however, and age profiles in the developing countries also will change as their proportions of children fall and those of older people rise; crude death rates then will escalate gradually.[2] Figure 2–1 shows the direction of mortality change from 1955 to the present and projects change to 2050 for both groups of countries.

Other Developed Countries

Since aging of the population affects death rates in developed countries, their mortality experiences can be compared by measuring the pace of aging, using an *index of aging* to rank ten countries and assessing variations in their crude death rates and age-adjusted death rates. The index of aging (Table 2–2) is the number of persons aged 65 and over per 100 persons aged 0–14.[3]

The United States has the lowest index of aging among the ten developed countries and its crude death rates for both sexes also are relatively low. Age standardization, however, shows that some other nations have lower age-adjusted death rates than the United States, partly because of significant mortality differentials among America's component groups. The difference between blacks and whites, especially black males and white females, makes for a less favorable overall mortality situation in the United States than in many other countries at comparable levels of industrialization and urbanization. The shorter average life expectancy among Hispanics than non-Hispanics has similar effects. These comparisons reflect the lower average level of living and lesser availability of health care still experienced by certain groups, in part because the United States is the only developed nation without a comprehensive health plan providing universal coverage. In contrast, Switzerland and Sweden have the lowest age-adjusted death rates because their proportions of disadvantaged persons are small and excellent health care is readily available.

Developing Countries

Table 2–2 also shows the index of aging, crude death rate, and age-adjusted death rate for ten developing nations. Their indexes of aging are much lower than in any developed country, thus reflecting their youthful age profiles. In fact, even the lowest index of aging among the developed countries is three times greater than the highest among the developing countries. Consequently, crude death rates are also much higher in the developed than the developing nations. Comparisons among regions and large categories of countries show a somewhat different pattern. Latin America, for example, has the lowest crude death rate, partly because of its youthful age structure, but also because mortality actually has fallen rapidly to low or moderate levels in much of that region (see Figure

Figure 2–1

Crude Death Rate, by Development Category, 1955–2050

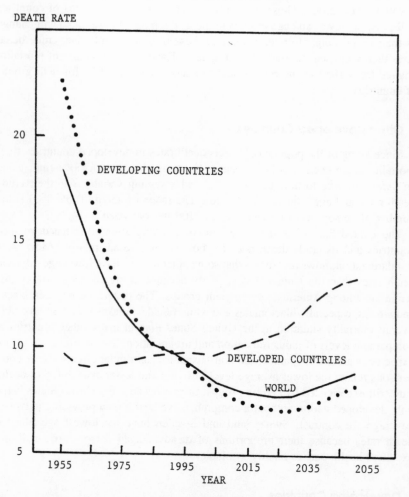

DEATH RATE

Source: U.S. Bureau of the Census, Ellen Jamison, Peter D. Johnson, and Richard A. Engels, *World Population Profile: 1987* (Washington, DC: Government Printing Office, 1987), Figure 14.

2–2). The Near East and North Africa also have relatively low rates, followed by developing Oceania and Asia. The developed countries come next because of their aging populations, even though their overall life expectancy is six years greater than in Latin America. Sub-Saharan Africa has the highest death rate— more than twice the Latin American average and 50 percent above the world figure.[4]

Age standardization changes the ranking of the developing and developed

Table 2-2
Index of Aging, Crude Death Rate, and Age-Adjusted Death Rate, Selected Countries, by Sex, 1990s

Country	Index of Aging[a]	Crude Death Rate		Age-adjusted Death Rate[b]	
		Male	Female	Male	Female
Developed countries					
Sweden	101.8	12.1	10.4	8.3	7.5
West Germany	98.0	11.5	11.6	10.0	9.1
Switzerland	85.9	9.8	8.6	8.2	7.0
Denmark	81.8	12.1	10.7	9.8	8.9
England and Wales	80.4	12.0	11.7	10.0	9.2
Austria	78.4	11.7	12.0	10.5	9.6
East Germany	70.4	12.6	14.4	12.0	11.9
Netherlands	62.1	9.2	7.8	9.1	7.7
France	60.0	10.7	9.4	9.8	7.9
United States	54.9	9.5	8.1	9.5	8.1
Developing countries					
Chile	18.0	6.8	5.4	11.4	10.6
Sri Lanka	12.3	7.1	5.0	10.9	11.6
Costa Rica	12.2	4.6	3.6	7.7	8.3
Panama	11.6	4.3	3.4	7.2	8.1
Suriname	11.2	9.0	6.7	13.9	11.3
Malaysia	9.6	6.1	4.6	11.9	12.1
Philippines	7.8	7.3	5.1	11.2	10.7
Mexico	7.1	7.1	5.3	11.6	11.2
Guatemala	6.4	9.4	8.0	14.1	15.4
Kuwait	3.1	2.7	2.3	9.5	10.7

Source: United Nations, Demographic Yearbook (New York: United Nations, various editions); National Center for Health Statistics, Vital Statistics of the United States, Vol. II, Mortality, Part A (Washington, DC: Government Printing Office, various editions).

[a]Number of persons aged 65 and over per 100 persons aged 0–14.

[b]Standardization based on U.S. age profile in the year for which age-specific death rates are available for each of the other countries

countries because the age-adjusted death rate is almost always higher in the former than the latter (see Table 2–2). There are some exceptions, however, because deaths are underreported and some so-called developing countries are actually fairly close to full development. They are well along the path to good general health, and even their age-specific death rates compare favorably with those in many fully developed countries. One such case is Costa Rica, where death rates are at or below those in the United States at many specific ages, especially late childhood, young adulthood, and middle age. Costa Rica has a high literacy rate and is extending its system of national health insurance and

Figure 2–2
Crude Death Rate, by Region, 1989

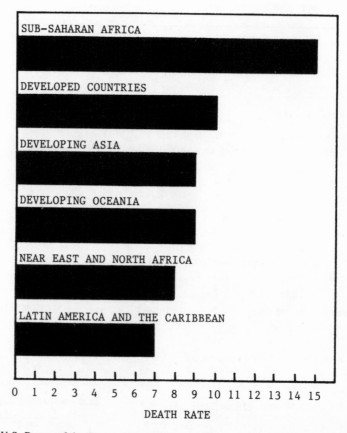

SUB-SAHARAN AFRICA

DEVELOPED COUNTRIES

DEVELOPING ASIA

DEVELOPING OCEANIA

NEAR EAST AND NORTH AFRICA

LATIN AMERICA AND THE CARIBBEAN

0 1 2 3 4 5 6 7 8 9 10 11 12 13 14 15

DEATH RATE

Source: U.S. Bureau of the Census, Ellen Jamison, *World Population Profile: 1989* (Washington, DC: Government Printing Office, 1989), Table 2.

old-age protection throughout the country.[5] It also lacks a subservient, disadvantaged minority population and has a high degree of social, racial, linguistic, and educational integration.[6] These conditions contribute directly and indirectly to Costa Rica's favorable mortality situation, even though its per capita GNP is only a tenth of that in the United States.

Even countries with many poor people and relatively high age-adjusted death rates have significantly lower rates than a few decades ago. In fact, rapidly falling death rates, especially infant mortality rates, combined with stubbornly high birth rates, caused most of the huge population increase in the developing countries after World War II. Except for a few seriously underdeveloped places, death rates have dropped dramatically in the developing nations, though considerable reduction is still possible; so is continued rapid population growth unless birth rates fall at about the same pace as death rates. Significant fertility reductions are occurring in many developing countries, but persons born when birth rates were high are large reproducing populations that contribute ever-growing numbers of children, despite decreasing reproduction per couple.

The percentage of all deaths concentrated in the first year of life also separates developing and developed nations, as does the proportion of all deaths among the elderly. Table 2–3 shows ten developing countries whose infant losses are relatively high and ten developed countries whose losses are quite low. In the developing countries collectively infant deaths account for 21 percent of all deaths among males and 24 percent among females, while in the developed countries the average is only 1.2 percent among males and 0.9 percent among females. Conversely, in the developing countries the deaths of men aged 65 and over account for only 26 percent of all male deaths, those of elderly women for just 32 percent of all female deaths, but in the developed countries the figures for men and women are 71 percent and 84 percent, respectively.

These differentials reflect the higher percentage of infants in the developing countries and the lower proportions of elderly people, but even accounting for those differences, deaths in the developing countries are still heavily concentrated in the first year. In Egypt, for instance, 25 percent of all male deaths and 27 percent of all female deaths occurred in the first year, even though only 1.9 percent of all males and 1.9 percent of all females were in that age group. In the Netherlands, on the other hand, a mere 1.2 percent of all male deaths and 1.1 percent of all female deaths struck infants. Since 1.3 percent of all males and 1.2 percent of all females were in their first year of life, the incidence of infant deaths is nearly in proportion to the percentage of infants.

The distribution of deaths among the elderly is also quite different in the developed and developing nations. In the Netherlands, for instance, 72 percent of all male deaths and 83 percent of those among females occurred in the ages 65 and over. In Egypt only 28 percent of the male deaths and 34 percent of the female deaths took place then because causes that are well controlled in most developed countries still take such a high toll of Egyptian infants and children that far fewer persons survive to 65. Elderly people are comparatively small percentages of the population in Egypt and other developing countries, however, so the lower concentration of deaths among the elderly is more in line with the age profile than is the heavy loss of infants. That loss reflects many disadvantages that produce poor health care, short life expectancy, and high risks to the vulnerable young.

DIFFERENTIALS BY SEX

Regardless of race or ethnicity, American males at every age have higher death rates than females, although the gap is narrower at some ages than others. Table 2–1 shows differentials in age-specific death rates, Figure 2–3 the *male/ female mortality ratio*, or the male death rate at specified ages as a percentage

Table 2–3
Percent of All Deaths Attributable to Persons Aged 0–1 and 65 and Over,
Selected Countries, by Sex, 1980s

Country	Age 0-1		Age 65 and Over	
	Male	Female	Male	Female
Developed countries				
United States	2.0	1.7	64.2	77.8
Netherlands	1.2	1.1	72.4	82.8
West Germany	0.9	0.6	69.7	85.8
France	1.3	1.0	66.9	84.7
Denmark	1.0	0.6	72.2	81.8
Austria	1.2	0.8	68.2	86.1
East Germany	1.2	0.7	68.3	85.7
Sweden	0.7	0.7	78.9	87.2
Switzerland	0.9	0.8	72.2	85.4
England and Wales	1.2	0.9	73.5	84.2
Developing countries				
Guatemala	26.8	25.8	20.7	23.3
Egypt	25.1	26.8	27.7	34.1
Zimbabwe	20.9	29.7	16.0	14.8
El Salvador	16.3	19.5	26.0	37.9
Bangladesh	30.6	30.6	22.9	16.5
Mexico	15.8	16.4	25.0	43.8
Pakistan	45.0	42.2	21.3	21.3
Sri Lanka	11.1	13.4	39.9	44.9
Kuwait	16.6	23.4	26.1	36.5
Venezuela	16.2	15.9	35.0	46.0

Source: United Nations, Demographic Yearbook, 1988 (New York: United
Nations, 1990), Table 19.

of the female death rate at those ages. Both data sets show a substantial mortality
gap by sex from birth onward, though it even appears among embryos and
fetuses.[7] In 1987, for example, 7.9 male fetuses (presumed gestation of 20 weeks
or more) died per 1,000 live births, compared with 7.4 females, although in
twin deliveries among blacks and other multiple deliveries among whites the
female fetuses were more likely to die.

The mortality differential by sex varies considerably with age. In the ages 0–
9 white males have a death rate about one-third again as high as white females;
black males have a rate about a quarter again as high as black females. The
male disadvantage then escalates rapidly, until in the ages 20–24 death rates
among males of both races are more than three times those among females. After
that peak, the gap narrows with age, until males 85 and over have death rates
only about one quarter again as high as those among females. At no age, though,
does the male/female differential disappear, let alone reverse.

The sex mortality differential is often thought to exist in all societies at all

Figure 2–3
Male/Female and Black/White Mortality Ratios, by Age, 1985–87

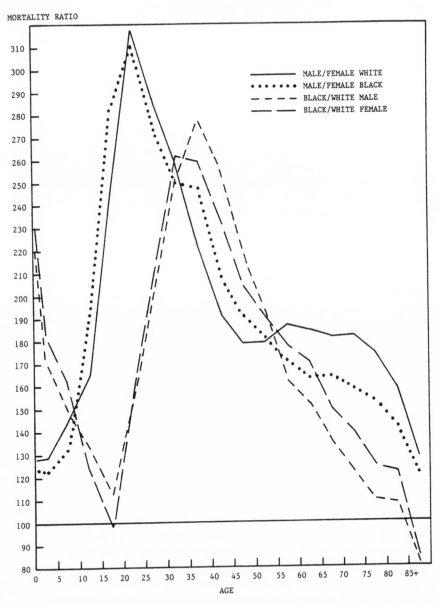

Source: National Center for Health Statistics, *Vital Statistics of the United States*, 1985, 1986, 1987, Vol. II, *Mortality*, Part A (Washington, DC: Government Printing Office, 1988–90), various tables.

ages because of biology. Francis Madigan, for example, underscored biological differences in his comparison of Roman Catholic male and female celibates subject to nearly identical social conditions.[8] But in comparing large national populations, George Stolnitz found numerous exceptions to the greater longevity of females because of high rates of maternal mortality, female infanticide, and other conditions that reflect low status and poor treatment of females in some societies.[9] Those departures from the U.S. pattern are closely connected with women's level of living, and in some societies where their status is inordinately low, the reported death rates of girls 1–4 and of very elderly women are higher than the rates of males in those ages. This is especially likely in some Middle Eastern and African Islamic populations, though due to poverty and underdevelopment, not Islam.[10] Despite the exceptions, the greater longevity of females is so pervasive that animal studies even show it to exist almost universally among other species.[11] Consequently, it cannot be attributed solely to environmental conditions, but also to a significant biological base. In fact, where environmental conditions are about the same for the sexes and especially where they improve markedly and overall mortality levels drop, the exceptions tend to disappear and females are more durable on the average at nearly every age.[12] Biologically, females seem to benefit from their two X chromosomes, compared with single ones in males; premenopausal women are protected by estrogen, especially from heart disease; women adjust better to changes in temperature; and females have other inherent attributes that help increase their longevity.[13]

Chapter 4 explores differential mortality by cause and Chapter 5 analyzes trends by cause, but here we note that efforts to explain the sex mortality difference, especially its rapid increase in developed countries during the twentieth century, must assume complex interactions between biological and environmental factors, particularly as reflected in causes of death. Otherwise, the biological advantage of females would never be canceled by variable social conditions.[14] In fact, social influences intensify the differential in the United States and other places where most women do not have unusual life-threatening risks, especially during pregnancy, childbearing, and old age. Women are more likely than men to seek medical attention for their illnesses, are less apt to be killed driving automobiles, and are still less likely to be heavy smokers and drinkers. Even now they seem better able to handle stressful situations or are less likely to be in those situations, though changes in women's occupational and other roles have altered this condition significantly.

Smoking patterns are particularly strategic: Robert Retherford suggested as early as 1972 that about half the mortality differential by sex could be explained by different rates of cigarette smoking, which produces high rates of respiratory cancer and circulatory disease.[15] Since then the death rate for lung cancer has risen faster among women than men because more women have become heavy or long-term smokers and the incidence of smoking has fallen faster among men than women. For example, the age-adjusted death rate for respiratory cancer among white males rose from 34.6 per 100,000 in 1960 to 58.0 in 1980 and then remained at about that level; in 1988 it was 68 percent higher than in 1960. Among white females the age-adjusted rate rose from 5.1 in 1960 to 18.2 in 1980 and continued to climb to 24.8 in 1988, when it was 386 percent higher than in 1960. In 1960 white men were about 6.8 times more likely than white women to die of respiratory cancer, but by 1988 men were only 2.3 times more likely to be its victims—the sex differential had narrowed by two-thirds. Furthermore, in 1960 white women were one-fifth as likely to die from lung cancer as from breast cancer, but by 1988 their age-adjusted death rates for both causes were about the same, even though the breast-cancer death rate also rose a bit.

The pattern is similar among blacks, although black women have about the same lung-cancer death rate as white women, while black men have a significantly higher rate than white men.[16] Women are still less likely than men to die of lung cancer, but their relative advantage has been slipping away. This is partly why the gap between the sexes in life expectancy at birth among whites fell from an all-time high of 7.8 years in 1975 to only 6.6 years in 1988; among blacks it fell from 8.9 to 8.5 years.[17] Smoking is implicated in these changes: between 1965 and 1987 the proportion of white men aged 18 and over who smoked cigarettes fell 40 percent, while the proportion among white women fell only 21 percent; in 1987 about 30 percent of the males and 27 percent of the females were smokers. This convergence helped shrink the sex mortality differential for respiratory cancer. Yet the incidence of cigarette smoking did drop for both sexes while the death rate for lung cancer rose, so other factors are involved. In addition, it takes a long time for smoking to produce fatal lung cancer, so the effects of recent decreases in smoking will be felt later, whereas current losses to lung cancer reflect smoking patterns that began much earlier.

Changes in the age-adjusted death rate for heart disease—the leading cause of death for both sexes—have had less effect on the male/female mortality differential. While both sexes experienced improvements between 1960 and 1988, the death rate among white males fell 41 percent and among white females 42 percent. Among blacks the differential decrease was somewhat greater: the age-adjusted rate fell 41 percent for males and 38 percent for females. On balance, the changes in death rates for heart disease preserved most of the sex mortality gap, even while changes in death rates for respiratory cancer decreased it somewhat. Change in mortality for heart disease and cancer illustrate the impact of specific causes of death on the sex mortality differential, but since both causes (and others discussed in Chapters 4 and 5) are affected by interactions between biology and environment, they also reflect the complexity of that differential.[18]

The sex mortality differential among whites also narrowed for chronic obstructive pulmonary diseases (e.g., emphysema, bronchitis), partly because of changes in smoking behavior. White males are still almost three times more likely than white females to die in motor vehicle accidents, but the death rates have fallen for both sexes and the sex differential was less in 1988 than in 1960 because the rate dropped faster for males.

The sex differential among whites grew between 1960 and 1988 for all other major causes of death, including diabetes, colorectal cancer, suicide, homicide, pneumonia and influenza, nonvehicular accidents, and stroke. Males already had higher death rates for those causes (except diabetes) and their disadvantage intensified.

Sex differentials by cause of death among blacks followed most of the same patterns, but the degree of change was generally greater for black people. Thus, the age-adjusted death rate for respiratory cancer rose much faster among black women than black men, so the sex gap for that cause narrowed substantially. The sex mortality differential among blacks narrowed for only one other cause—motor vehicle accidents—though males are still more than three times as likely as females to die by that means. The sex differentials widened among blacks for all other major causes, and in every case the pace of increase was greater than among whites.

Sex Differentials by Age

So far we have used the age-adjusted death rate to examine the sex mortality differential. That index accounts for changes in the age structure of the population but does not show changes in the importance of causes of death at particular ages. Thus, we can look at age-specific death rates, illustrated by changes in respiratory cancer between 1959–61 and 1985–87. The largest increase in the death rate for respiratory cancer among women occurred in ages 55–74, while for men they occurred in the older ages (see Table 2–4). In fact, the rate even fell 11 percent for white men aged 35–44. At every age, however, the increase was much greater for white women than white men, though among blacks aged 75 and over it was greater for men than women. At most ages, therefore, the sex mortality differential for lung cancer decreased.

These variations reflect differences in smoking behavior by age, including the length of time a person has smoked and whether he/she is still smoking or has quit. The prevalence of cigarette smoking does not vary much among the age groups under 55, though it reaches its peak in the ages 35–44. By 55–74, however, the prevalence of smoking drops off, largely because about a third of those persons who were smokers have quit. The low prevalence among persons past age 75 is mostly because they never started.[19] But the effects of smoking are long-lasting, and many persons who smoked for two or three decades and then quit have already incurred the damage that pushes up the respiratory cancer death rate with age. Indeed, that rate jumps among persons in their 50s and 60s and continues to rise into the 70s. At the same time, reduced rates among elderly females underscore both the health advantage enjoyed by lifetime nonsmokers and the deaths of many long-time smokers before reaching the oldest ages. Among the nonsmokers, very elderly women are part of the generation for which cigarettes were a forbidden vice, and smaller percentages began using them when they were young. Whatever its degree of attraction, smoking has the most important effect on different rates of change in mortality for respiratory cancer in

Table 2–4
Percent Change in Age-Specific Death Rate for Respiratory Cancer, by Race and Sex, 1958–60 to 1985–87

Age	White		Black	
	Male	Female	Male	Female
35-44	-10.9	+78.1	+16.8	+110.0
45-54	+24.7	+294.6	+97.7	+248.4
55-64	+52.3	+490.7	+142.3	+433.0
65-74	+96.8	+484.2	+236.4	+425.7
75-84	+218.1	+288.4	+391.8	+317.6
85 and over	+267.2	+155.0	+331.3	+288.2

Source: Robert D. Grove and Alice M. Hetzel, Vital Statistics Rates in the United States, 1940-1960 (Washington, DC: Government Printing Office, 1968), Table 63; National Center for Health Statistics, Vital Statistics of the United States, 1985, 1986, 1987, Vol. II, Mortality, Part A, (Washington, DC: Government Printing Office, 1988-90), various tables.

the several age categories. More broadly, "cigarette smoking is the single factor responsible for most preventable deaths in our society."[20] These excess deaths among smokers come not only from lung cancer, but from coronary heart disease, other cardiovascular diseases, chronic obstructive pulmonary diseases, and cancer of the larynx, upper digestive system, bladder, pancreas, and kidneys.[21] Altogether, smoking is probably responsible for over 400,000 deaths annually.

Smoking is not the only reason people die from lung cancer. Escaping radon gas, for example, is thought to cause as many as 20,000 lung cancer deaths a year, or about 16 percent of the total. The most important influence may be genes internal to cells that turn normal cells into cancerous ones, often after the genes have been damaged by various carcinogens.[22] Cigarettes, however, are the principal carcinogen that causes gene damage and lung cancer.

Age and sex differentials in mortality for respiratory cancer and other causes of death point up the complex interplay between biological factors and the prenatal and postnatal environment. While one's intrinsic life span depends on genetic make-up, life is never lived in an ideal environment and the inevitable difference between intrinsic life span and actual length of life "will be determined by interactions among the person's genetic endowment, stage of development as indicated by age, and the quality of the environment over time."[23]

The role of genetics in longevity is more difficult to identify and manipulate than environmental factors, but biomedical research is providing more clues about specific genes for colorectal cancer, breast cancer, Alzheimer's disease, and other causes of death. In some cases the first hint of a genetic component appears as high rates of certain diseases in families; and more sophisticated research in genetics provides the tools to investigate those patterns. More patterns seem due not just to similar lifestyle, but to intergenerational genetic make-up and inheritance. Thus, we are closer than ever to identifying specific genes implicated in certain diseases, learning what makes them abnormal, and discovering how they exert their carcinogenic and other pathological effects. In turn, that brings us closer to practical screening procedures for genetic aberrations and to new interventions that will halt or slow the effects of defective genes and prevent certain diseases.

Mortality and Morbidity

Mortality variations between men and women differ from variations in *morbidity* (disease). Although females have lower death rates and higher life expectancy at every age, they are more likely to seek care for illness. Consequently, they are more likely to report suffering acute and chronic illnesses, but often their problems are less life threatening than those among men.[24] For example, the National Health Inventory Survey (NHIS) shows that women, especially older ones, report significantly higher rates of illness from acute conditions such as respiratory diseases, afflictions of the digestive system, ear infections, and headache.[25] They are also more likely to suffer certain chronic maladies such as arthritis, hypertension, varicose veins, and allergies—conditions that either are not inherently life threatening or can be controlled.[26] Women are also more likely than men to suffer short-term disability and to use medical services and medication when ill.[27] For example, women in specific age groups have higher rates than men of restricted-activity days because of acute conditions. They also have higher rates of bed days because of those conditions; employed women have a greater average number of work-loss days than employed men; females aged 5–17 have higher rates of school-loss days than their male counterparts.[28]

These variations involve not just biological but environmental differences between the sexes, including risk avoidance, personal attentiveness to health, and other behavioral expectations still attached to gender roles. Men who suffer influenza, for example, are less likely than female sufferers to spend time nursing the illness in bed and are, therefore, less likely to incur short-term disability, even though their infections may be just as severe. Women are more likely to have health knowledge and to seek various forms of assistance when they are ill.[29] While such efforts may not cure an illness, they do represent intervention to forestall death. In short, both biological and environmental factors "result in different disease risks and illness experiences."[30] Consequently, part of the sex mortality differential is due to the kinds of illnesses women suffer and their attention to health and amelioration, even though women's overall incidence of reported illness is greater than men's. On balance, however, men have fewer years of good health, while "females live longer and their longer life is composed of more years free of disability, more years with both short- and long-term disability, and more years in institutions."[31]

DIFFERENTIALS BY RACE AND HISPANIC ORIGIN

Blacks and Whites

Virtually every comparison shows that whites have significantly lower death rates than blacks. That was true in 1900, when the first life tables showed life expectancy at birth to be only 32.5 years for black males but 46.6 years for white males, and 33.5 years for black females but 48.7 years for white females. The racial differential persists, and though the gap has narrowed, in 1988 the racial variation in life expectancy was still 7.4 years for males and 5.5 years for females. In that year the crude death rate was higher for black males (10.1) than white males (9.5) and lower for black females (7.5) than white females (8.6), but only because a larger percentage of blacks is in younger age groups where death rates are low. Therefore, further consideration of differential mortality by race needs age controls.

The age-adjusted death rate among white males in 1988 was 6.6, compared with 10.4 among black males; among females the age-adjusted rates were 3.8 and 5.9, respectively. Blacks are at a disadvantage across most of the age spectrum, except 15–19 and 85 and over, when a much-studied *mortality crossover* seems to occur (see Table 2–1). Figure 2–3 shows the black/white mortality ratio by sex (black age-specific death rate as a percentage of white age-specific death rate) and further illustrates the racial differentials by age. The death rate among black infants is more than twice that among white infants, but the black/white mortality differential then declines steadily among both sexes until ages 15–19. That is when the death rate for motor vehicle accidents among whites is more than twice that among blacks, and the higher death rates among blacks from other causes are not enough to offset fully the large discrepancy in motor vehicle deaths. By age 20, however, blacks again have a large mortality disadvantage, partly because of homicide, and from age 30 to 49 their death rates are at least twice those of whites, even though death rates for all races rise significantly with age. Moreover, at most ages the black/white mortality ratio is higher for females than males because white females enjoy greater longevity than black females and males of both races.[32]

After middle age the racial differential declines substantially, and the crossover occurs among people in their 80s.[33] Part of this reported phenomenon is due to data inaccuracies at the oldest ages, especially for blacks, and base census figures and reported deaths contain enough errors to account for some of the crossover.[34] Medicare data, however, substantiate the crossover, though they are not specific for cause of death and go back only to 1966.

Data inaccuracies notwithstanding, blacks who make it to the oldest ages despite environmental stress and relatively high death rates for most causes do seem "destined by natural selection to live an especially long life."[35] Differences between races as a whole in education, income, health care, and other aspects of socioeconomic status produce higher death rates among blacks in the younger ages that winnow out the least durable individuals, leaving less variation in the health conditions and survival potential among the oldest blacks than the oldest whites. This is reflected not only in the crossover in the oldest ages, but also in the gradual narrowing of the black/white mortality gap after age 45 or so, when the aging process and morbidity become more alike for the races, until elderly blacks finally gain the advantage. Since the crossover has roots in racial mortality variations in the younger ages, the crossover actually reflects three conditions: (1) greater social, economic, and health disadvantages for blacks than for whites; (2) markedly higher age-specific death rates for blacks than for whites through middle age, essentially because of the disadvantages; and (3) a slower rise in the death rate of blacks than whites after middle age, until the two rates converge in old age and the convergence finally produces the crossover.[36]

If blacks are less advantaged as a whole than whites, why is the rate of increase in age-specific rates slower for blacks after middle age? How can convergence and crossover be explained? In general, blacks seem more likely to die directly from certain degenerative conditions, while whites tend to live longer even while afflicted by them. But those degenerative conditions also kill white people, though less directly and at an older average age because of better medical treatment. Thus, even though conditions are worse for blacks in the young and middle ages, survivorship selects some with the stamina to outlive whites. The apparent advantage of whites is actually a consequence of treatment and other intervention and they, therefore, are more apt to succumb in their 70s or early 80s, while their black counterparts survive longer.[37] A white person who has lived to 80, for example, may be more frail or more likely to live alone than a black person of the same age and sex, and the white person may die sooner than the black person for these and other reasons.

It is easy to overgeneralize this Darwinian notion of selection and to miss other possible explanations, especially differential mortality by cause, discussed in Chapters 4 and 5. Nonetheless, convergence and crossover in age-specific black and white mortality rates seem real. In addition, convergence and crossover reflect other similarities among persons who survive to the oldest ages—similarities that are greater than at birth and in the 30s and 40s, when the black/white mortality discrepancy is largest.[38]

Racial variations in the proportion of elderly persons who live alone may affect the crossover, especially if solitude worsens health care, nutrition, loneliness, and despair. Higher proportions of elderly black than white women live with relatives or nonrelatives instead of alone, and that may compensate for other disadvantages. Elderly black men, however, are more likely than whites to live alone. In 1988, for example, 54 percent of all white women 75 and over but only 45 percent of black women lived alone; among men the proportions were 30 percent for blacks and 23 percent for whites.[39] The higher incidence of family or quasi-family living arrangements among black women probably contributes to their greater longevity and to the mortality crossover. This does not seem to apply to black men in quite the same way, though they are more likely than elderly white men to maintain kinship and friendship ties, probably with similar constraints on mortality.

The crossover is affected by rates of morbidity and functional disability and by high-risk behavior such as heavy smoking and drinking: when these are held constant for the races, the magnitude of the crossover diminishes. Lisa Berkman and colleagues also found that type of housing significantly affects the crossover: in public housing healthy blacks tend to outlive healthy whites, while in other residential arrangements the reverse is true.[40] These findings suggest that many elderly blacks not in public housing live under various adverse conditions that shorten life. Within public housing elderly blacks are often part of a close-knit community of friends and relatives who contribute to longevity, however indirectly, whereas whites are more likely to be isolated, to be in public housing because of downward mobility, and to perceive that situation negatively and their relationships as less supportive and even threatening.[41] The findings also reflect great diversity in the elderly population, not just by race, but by socioeconomic status, morbidity, high-risk behavior, environment, access to health care, and other variables. Consequently, conclusions drawn from aggregate data—in this case about the racial crossover—must be tempered by an understanding of variations among individuals and subgroups, some subject to the crossover, others not.[42]

The longevity advantage that women have over men shows up in the racial comparisons long before the racial crossover. The death rate of black females, which is substantially below that of black males at every age, particularly 15–44, also falls below that of white males in the ages 10–14 and remains there until 25–29 (see Table 2–1). This advantage of black females over white males then disappears but reappears in all ages 55 and over. Thus, while race differentials in age-specific death rates have decreased over the long run, sex differentials in both races have increased substantially, though with some exceptions in the 1980s. Consequently, in 1988 life expectancy at birth of black females exceeded that of white males by only 1.1 years but it was 5.5 years behind that of white females.[43] Significant discrepancies between the races will persist as long as average socioeconomic levels differ, but the long-term racial convergence in death rates is an important improvement. It is also reflected in morbidity rates, use of health care, levels and periods of disability, and other indexes. The pace of improvement slowed in the 1980s, however, and there were enough reversals among poor groups that in certain respects the mortality gap between blacks and whites widened once more. The increase stems partly from poor medical care, both preventive and curative, among a large number of blacks, and from risky lifestyles.[44]

Other Races

The data do not enable a comprehensive analysis of mortality differentials among other races than blacks and whites. "Other races" identified by the Census

Bureau include persons whose heritage is Native American, Eskimo, Aleut, Chinese, Filipino, Japanese, Asian Indian, Korean, Vietnamese, Hawaiian, Samoan, and Guamanian.[45] The NCHS provides very few mortality data about these "races," partly because cross-classification by age and sex creates such small categories that the data are unreliable. In addition, death certificates supply data on only five racial groups: black, American Indian and Alaskan native, white, Chinese, and Japanese. The NCHS does provide many materials for "all other" as a residual category apart from blacks and whites, but those data yield little insight into the mortality situation of specific races. Even when the number of deaths is reported for them, the population counts or estimates necessary to compute rates, especially age-specific and age-adjusted rates, are generally available only for a census year, and even then they have limitations.

Information published in 1980 does show that Native Americans aged 0–4 have a death rate about two-thirds again as high as the one for whites. It rises to about 3.5 times higher in the ages 25–34 and then the rates converge, until after 65 or so a crossover occurs similar to that found among blacks, probably for similar reasons, including data flaws. The age-adjusted death rate computed from these age-specific rates is considerably higher among Native Americans than among whites, but lower than among blacks.[46]

Among persons of Chinese and Japanese ancestry age-specific rates are far below those of whites, sometimes less than half. As a result the age-adjusted death rate among Chinese is only 73 percent of that among whites and the rate among Japanese is only 49 percent. Except for Native Americans, Chinese are the largest population of other races, and their age-adjusted death rate and all age-specific rates are lower than those among whites, although the death rates for both groups are fairly high during infancy, decrease with age until about 15, and increase steadily thereafter.[47] Moreover, age-adjusted sex-mortality ratios for the two races are similar, with females having a distinct advantage, although the sex ratio is higher for whites at some ages (15–55), for Chinese at other ages (75 and over).[48] The four leading causes of death, in descending order, are the same among whites and Chinese: heart disease, cancer, cerebrovascular diseases, and accidents. The fifth cause, however, is chronic obstructive pulmonary diseases among whites and pneumonia and influenza among Chinese. Suicide ranks tenth among whites but seventh among Chinese; the death rate for that cause is particularly high among older women, especially those born in China. In addition, heart disease, accidents, chronic obstructive pulmonary diseases, chronic liver disease and cirrhosis, and atherosclerosis are responsible for higher percentages of deaths among whites than Chinese, while the reverse is true for cancer, pneumonia and influenza, diabetes, and suicide.[49] Overall, however, Chinese Americans are healthier and enjoy greater longevity than white Americans.[50]

Hispanics and Non-Hispanics

The number of age-specific deaths among Spanish-origin people was first reported by the NCHS in 1984 for the Hispanic population as a whole and for

Mexican, Puerto Rican, Cuban, and Central and South American subcategories. The NCHS does not report age-specific or age-adjusted death rates for Hispanics or the subgroups, however, and computing them, using estimates of the Hispanic population by age in the *Current Population Reports*, is difficult methodologically. In particular, one needs caution in using data on number of deaths from one source (NCHS) and population estimates from another (Census Bureau sample).[51] Furthermore, in 1988 the mortality data for Hispanics were reported by only 26 states and the District of Columbia, and while those data were 90 percent complete, they were based on place of occurrence, not residence.[52] At the same time, the Census Bureau does not report the age data by states necessary to compute age-specific or age-adjusted death rates for the same 26 states that provide death statistics. For these reasons, few mortality data on the Hispanic population are reported in this chapter, though some conclusions are possible.

In first-generation Hispanic groups, mortality levels are relatively high among adolescents and young adults, especially males, because of deaths from violence and, increasingly, HIV/AIDS. On the other hand, elderly Hispanics of the first generation have significantly lower death rates than elderly non-Hispanics for heart disease and cancer. These relationships apply to persons of Mexican, Cuban, and Puerto Rican origins, and patterns at various ages tend to offset each other, so the overall mortality level of Hispanics is not very different from that of non-Hispanic whites.[53] Even so, there are important local variations, one of which causes Puerto Ricans living in New York City to have higher death rates than non-Hispanic whites, while Puerto Ricans living elsewhere in the United States have much lower death rates than their New York counterparts.[54] The difference is reflected in cause of death: those attributable to lifestyle (e.g., homicide, cirrhosis of the liver) underscore the extent to which many mortality variations result from local conditions. In this case "local" means not just New York City as a whole, but specific parts of the city that contain a large share of the Puerto Rican population, especially the poor segment. Among the New York City group average family income and educational levels are lower than among Puerto Ricans living outside the city, and the disparities help produce different mortality patterns; so do different rates of violence and substance abuse. Other Puerto Ricans are concentrated mostly in the urban parts of New Jersey, Illinois, Florida, California, and Pennsylvania, and their mortality rates are well below those of New York City residents.[55]

DIFFERENTIALS BY MARITAL STATUS

Married Americans have lower mortality and morbidity rates than those who are single, widowed, and divorced, regardless of age, sex, and race.[56] Furthermore, at most ages married men seem to have a greater advantage over men in the three unmarried categories than married women have over unmarried women, though with some exceptions.[57]

The patterns for 1985–87 appear in Table 2–5. Mortality data by marital status

Table 2–5

Age-Specific Death Rate, by Marital Status, Race, and Sex, 1985–87

Race, Sex, and Age	Single[a]	Married[a]	Widowed[a]	Divorced[a]
White male				
25–34	2.6	1.0	13.0	3.4
35–44	6.3	1.7	7.8	5.3
45–54	12.0	4.8	11.6	13.3
55–64	24.1	13.6	31.2	32.1
65–74	48.9	33.3	53.6	62.3
75 and over	140.7	91.3	164.0	144.9
White female				
25–34	1.0	0.4	1.9	1.1
35–44	2.4	1.0	2.6	2.0
45–54	5.7	2.8	5.5	4.9
55–64	13.4	7.1	12.2	12.2
65–74	27.8	16.7	24.3	28.4
75 and over	102.2	52.3	99.5	120.2
Black male				
25–34	5.4	2.4	14.2	4.7
35–44	13.7	4.4	19.9	8.9
45–54	26.7	9.3	25.2	19.8
55–64	37.7	21.9	37.4	33.1
65–74	82.7	41.3	62.4	80.5
75 and over	142.7	96.0	127.0	133.1
Black female				
25–34	1.9	1.0	2.8	1.6
35–44	4.6	2.2	5.5	3.0
45–54	10.1	5.6	8.4	6.5
55–64	20.9	11.7	20.0	14.2
65–74	34.9	22.3	35.4	27.4
75 and over	91.5	51.2	91.7	77.3

Source: U.S. Bureau of the Census, Arlene F. Saluter, "Marital Status and Living Arrangements," 1985, 1986, 1987, Current Population Reports, Series P-20 (Washington, DC: Government Printing Office, 1986-88), Table 1; National Center for Health Statistics, Vital Statistics of the United States, 1985, 1986, 1987, Vol. II, Mortality, Part A (Washington, DC: Government Printing Office, 1988-90), various tables.

[a]Deaths per 1,000 population in specified group.

are available for all groups 15 and over, but several marital categories contain so few persons aged 15–24 that the rates are suspect; Table 2–5 uses six age groups 25–34 . . . 75 and over. At every age the death rates of unmarried people are significantly higher than those of married people. In the unmarried group single people at some ages have lower rates than widowed and divorced people, while at other ages different combinations occur. Moreover, these greater risks among the unmarried have been increasing, not just in the United States, but in several other countries as well.[58] Walter Gove's 1973 conclusion that the mortality difference between being married and unmarried was greater for men than women also held in 1985–87 among both blacks and whites. At no age, however, did marriage push the death rate among men below that among women, so the familiar sex differential persists regardless of marital status, age, or race. Among white men, widowers have the highest mortality risks until about age 45 and again at age 75 and over, but in the ages 45–74 divorced men are in greatest jeopardy. Among white women, widows run the greatest risk of death until age 45, but from then on single and divorced women have higher death rates. Patterns similar to those among whites appear among blacks, except that single black men aged 45 and over are at greater risk than those widowed and divorced. After age 45, the risks among black divorced women are somewhat smaller than among single women or widows, although the difference between the last two categories is minor.

Why the striking mortality differences between married and unmarried people? One explanation is that on the average the least healthy people are also the poorest candidates for marriage; they may be less attractive to prospective partners than healthier people or be relatively isolated from social contacts by their disabilities. The most unhealthy and least stable people also seem less likely to remain married. These forms of selectivity do play a role, but a minor one.[59]

Other factors are more strongly implicated, especially how being married contributes to well-being and diminishes the significance of certain causes of death related to the psychosocial environment. Conversely, many unmarried people are relatively isolated from the close interpersonal ties that contribute to well-being, such as health care by a spouse.[60] Moreover, married persons report higher levels of happiness than the unmarried, so the causes of death most affected by ''one's psychological state'' create significant mortality differences between married and unmarried people.[61] For example, single, widowed, and divorced men have much higher death rates than married men for suicide, homicide, motor vehicle accidents, pedestrian deaths, and other accidents—all causes that involve various kinds of overt social acts.[62] The same is true among women, though the mortality differences for these causes are greater for widowed and divorced women than for singles.

Mortality from substance abuse shows a similar pattern, and unmarried men are far more likely than married ones to die from cirrhosis of the liver because of alcohol abuse, from lung cancer due to smoking, and from drug abuse. The same is true for women, but the differences are less than among men.[63] Married persons with illnesses that require prolonged care, such as tuberculosis and diabetes, also have significantly lower mortality rates for those causes than the unmarried, whether male or female. Finally, death rates for causes that are largely unaffected by social factors are about the same for married and unmarried persons.[64] Despite the possible effects of other arrangements, such as well-being of persons who are single but living with someone else, married persons have significantly lower mortality rates than single, widowed, and divorced persons.

The racial mortality crossover in the older years also appears in some of the marital-status categories, although the data are based on the whole group 75 and over, whereas in some sex and marital categories the crossover would not take place until age 80 or 85. Nevertheless, it appears in the ages 75 and over among widowed and divorced men and among women in all four marital categories. It does not show up for single or married men, but even in those cases the racial mortality ratio has converged significantly by age 75 and seems to be moving toward crossover.

Age-adjusted health data gathered in the National Health Interview Survey indicate that not only do married men have a lower death rate than unmarried men, but they are also less likely to experience activity limitation (15 percent) than are men who are single (20 percent), separated or divorced (20 percent), and widowed (19 percent). The relationship is similar but even stronger for women. Married persons of both sexes are also least likely to report being in only fair or poor health, while widows and widowers and separated women are most likely to so classify themselves.[65]

Women have a higher rate of restricted-activity days in each marital category, but married women and men have much lower rates than persons in the other three marital categories. Widows have the highest rate of restricted activity, followed by separated or divorced women and by singles. Separated or divorced men have the highest restricted-activity rate, followed by widowers and singles. Similar patterns by marital status and sex apply to the average number of days persons are confined to bed because of illness.[66] Thus, not only are mortality rates lower for married than unmarried people, but so are levels of morbidity and the constraints it imposes.

DIFFERENTIALS BY SOCIOECONOMIC STATUS

Mortality comparisons by race and other characteristics show how strongly socioeconomic factors are implicated in the differentials. Level of education seems a more strategic factor than income, occupation, and other indicators of socioeconomic status, partly because education is a fairly fixed index of status for most persons after age 25 or so, whereas the other factors often vary substantially throughout adulthood. Income level reported in a given survey, for example, may not reflect an individual's actual degree of affluence.[67] In addition, every adult has an identifiable level of education, whereas many persons, especially those 65 and over, do not have formal occupations or incomes that reflect their actual economic well-being. Unfortunately, however, the data on age-specific mortality by any of the socioeconomic variables, including education, are quite limited and enable only a brief look at their influence.

Education

In general, matched census and death certificate data show a strong inverse relationship between level of schooling attained and age-specific death rates, though it is less consistent for elderly persons than for those aged 25–64. This inverse relationship also appears among blacks and whites and men and women; it is especially marked for women aged 25–64 who have less than five years of schooling.[68] Furthermore, when Social Security Administration data, derived from individual files containing information on date of birth (i.e., age), race, sex, and date of death, are matched with data in the Current Population Survey, similar inverse patterns of mortality by level of schooling appear.[69] If we use the death rate for white men aged 25–64 at all levels of schooling collectively as a base (100), then any figure above 100 indicates a less favorable mortality situation, any figure below 100 a more favorable one. Harriet Duleep found that men with 0–11 years of schooling had a ratio of 119, but that high school graduates had one of 88 and men with at least a year of college one of 82, so the inverse relationship between level of education and mortality holds up.[70]

Levels of education among the elderly are particularly important, since deaths are heavily concentrated in that age group and rising levels of education should help lower their mortality rates, even though the inverse relationship is not as strong in their age group as in younger ones. In 1987, persons 75 and over still had significantly lower median levels of schooling than younger adults: elderly men had completed 9.6 years of schooling compared with 12.7 years for men aged 25–74; elderly women had completed 10.6 years compared with 12.6 for women aged 25–74.[71] In 1987 men 75 and over were 4 percent of the total male population, but they accounted for 36 percent of all male deaths; women 75 and over were 6 percent of the total female population but they accounted for 58 percent of all female deaths. Thus, mortality is heavily concentrated among persons with educational levels well below average for every five-year age group in the 25–74 category. Consequently, as better-educated cohorts enter the older years, their death rates should fall because better health and other advantages accompany higher levels of schooling. In fact, the significant drop in age-specific death rates of the elderly after 1968 owes a great deal to the rising average level of education among that age group. They are better informed about health, nutrition, exercise, and other factors that affect longevity, though the impact of education is not that simple, since education affects the ability to afford medical care, preretirement occupational status, and other strategic variables that also influence mortality levels.

Income

Age-specific death rates also fall as income rises, though this inverse relationship is closely tied to the one between mortality and education. Educational attainment helps determine income, with some variations, and people with higher income are best able to afford good nutrition, adequate housing, health insurance premiums, and other amenities that sustain health and prolong life. Thus, using the mortality rate of white men aged 25–64 at all income levels collectively as 100, Duleep found that persons with incomes of $2,000–$3,000 (in 1959 constant dollars) had a mortality ratio of 179 but that the figure dropped steadily as income rose, reaching 71 for those with $10,000 or more.[72] The pattern is roughly similar to the one found by Evelyn Kitagawa and Philip Hauser.[73] Improvements in income among persons 65 and over have been especially strategic in raising their life expectancy, and better Social Security coverage and cost-of-living increases helped account for the renewed downward trend in the elderly death rate after the mid-1960s. The enactment of Medicare on July 1, 1965, was an important cause of that trend, for between 1965 and 1970 health expenditures by people aged 65 and over almost doubled, while public payment of those costs rose from 30 to 61 percent.[74] The very old were most affected, and during that five-year period the death rate of males aged 85 and over fell 16 percent, while the rate among females dropped 21 percent—significantly faster than in any other age category, including infants. Thus, higher income helps promote lower death rates, whether the "income" is wages, entitlements, or some other form, and whether or not it appears in official income statistics.

Caution is in order, however, in interpreting the relationship between income and mortality among individuals. Suppose, for example, that a given person is ill and so cannot work full time; suppose also that the illness raises the probability of dying. These conditions would produce a correlation between mortality and income, but it is due to illness rather than to a causal relation between income and mortality. Establishing such causal relations is subject to less error when income and mortality are compared at the community level or in other aggregations rather than at the individual level. For that reason the preceding consideration of the connections between income and mortality levels is based on such aggregate data.

Occupation

Death rates are often related directly or indirectly to occupation. The relationships are difficult to document and measure, however, partly because the necessary data are not included in traditional compilations, partly because occupation and income have inseparable effects on mortality, and partly because occupations that minimize certain risks may intensify others. For example, some jobs that require regular physical exertion may be unusually prone to accidents but the exercise may retard arterial and heart problems, whereas jobs in which accidents are rare may be too sedentary. Some occupations produce relatively high death rates from specific causes, such as "black lung" in coal mining and "brown lung" in textile fabricating, and even many workers who live to retire have serious chronic illnesses and abnormally short life expectancies. Other occupations, such as logging, mining, and construction, are unusually subject to fatal accidents, and some accident survivors may succumb prematurely to disabilities incurred earlier. In general, however, the lowest mortality rates occur among professional and technical workers, the highest among laborers of various kinds; other categories are intermediate.[75] While the occupations themselves may play some part, they are less directly related to mortality rates than are the levels of living they provide. In fact, many apparent occupational differentials in mortality can be explained by other socioeconomic factors that affect the type of job one gets and the amount of income and health benefits it provides. Nonetheless, occupation and other elements of socioeconomic status are related to mortality, though education seems more closely related than any other status index. Judging from all three indexes, however, mortality differentials between persons who rank high and low on education, income, and occupation seem to be increasing, so the gap by social class seems to be growing.[76]

GEOGRAPHIC DISTRIBUTION OF MORTALITY

Geographic differences in socioeconomic status and other demographic characteristics help account for the distribution of mortality rates across the nation, and, in fact, the geographic distribution of poverty corresponds closely to the distribution of standardized death rates.[77] This section presents mortality variations by major region, division, and state, separately by sex.

Variations by Region and Division

Differences among the four regions in age-adjusted death rates of males and females are relatively small, partly because the regions are large units and mortality and many other demographic features are relatively homogenized, unless one sorts by race, class, and other characteristics related to the level of well-being. According to death rates among males, for example, the South fares the worst and the West the best, but the difference between them is only 1.0 death per 1,000 and the rate in the South is 15 percent higher than in the West (see Table 2–6). Among females the South and the Northeast have the highest rate and the West the lowest, but that difference is just 0.4 deaths per 1,000, or 11 percent. The variations are greater among the census divisions. Males have the highest age-adjusted rate in the East South Central division, the lowest in the Pacific and Mountain divisions, but the difference is still only 1.4 deaths per 1,000, or 22 percent. Females also have the highest rate in the East South Central division and the lowest in the West North Central division, for a difference of 22 percent. Thus, the nation's major subdivisions with the poorest mortality record are about one-fifth again as badly off as those with the best record.

Differential mortality by sex within the regions and divisions is another story, though a familiar one. The age-adjusted death rate among males ranges from 167 percent of that among females in the Pacific division to 181 percent in the West North Central division. Consequently, geographic differences pale beside

Table 2–6

Crude Death Rate and Age-Adjusted Death Rate, by Sex, and Sex Mortality
Ratio, Region and Division, 1987

Geographic Unit	Crude Death Rate		Age-Adjusted Death Rate		Male/Female Mortality Ratio[a]
	Male	Female	Male	Female	
United States	9.3	8.1	7.0	4.0	175.0
Northeast	10.3	9.2	7.2	4.2	171.4
New England	9.5	9.1	6.6	3.9	169.2
Middle Atlantic	10.6	9.3	7.4	4.2	176.2
Midwest	9.8	8.5	6.9	4.0	172.5
East North Central	9.6	8.4	7.2	4.2	171.4
West North Central	9.7	8.7	6.5	3.6	180.6
South	9.6	7.8	7.5	4.2	178.6
South Atlantic	10.1	8.2	7.4	4.2	176.2
East South Central	10.3	8.4	7.9	4.4	179.5
West South Central	8.6	7.0	7.3	4.1	178.0
West	7.9	6.8	6.5	3.8	171.1
Mountain	7.7	6.2	6.5	3.8	171.1
Pacific	8.0	7.0	6.5	3.9	166.7

Source: U.S. Bureau of the Census, Sixteenth Census of the United States:
1940, Characteristics of the Population, United States Summary
(Washington, DC: Government Printing Office, 1943), Table 7; Bureau of the
Census, "State Population and Household Estimates, With Age, Sex, and
Components of Change: 1981-88," Current Population Reports, Series P-25,
No. 1044 (Washington, DC: Government Printing Office, 1989), Tables 7, 8;
National Center for Health Statistics, Vital Statistics of the United
States, 1987, Vol. II, Mortality, Part B (Washington, DC: Government
Printing Office, 1989), Sec. 8, Table 6.

[a]Age-adjusted death rate of males as a percentage of age-adjusted death
rate of females.

large sex differences. In fact, they are so persistent throughout the nation that
in all regions and divisions the death rate of males is significantly higher than
that of females in every age category used to compute the age-adjusted rate. The
smallest sex mortality ratios occur in the youngest and oldest ages, but even
they are 125 or more in all divisions except New England and parts of the South.
Sex differences soar after early childhood, and by 15–24 age-specific death rates
among males throughout the nation are up to three times higher than those among
females. The ratios taper off gradually in the older ages, until the sex mortality
ratio among persons 85 and over is only about 125 or 130. No matter what the
geographic area, the sex differential is substantial for very young and very old
people and huge at most intermediate ages.

Variations by State

The smaller the geographic unit available for analysis, the larger the mortality variations are likely to be, so the county would be ideal to show those variations. In many counties, however, the number of deaths and base populations are so small that age-specific and age-adjusted death rates, computed separately for males and females, are not accurate. At the same time, crude death rates are misleading because of differences in age structures. Therefore, sex-specific age-adjusted death rates for the states best reflect detailed geographic mortality patterns (see Figure 2–4).

Age-adjusted rates among males range from 10.8 deaths per 1,000 in the District of Columbia to 5.0 in Hawaii, for a mortality ratio of 216. Rates among females range from 5.7 in the District of Columbia to 3.1 in North Dakota, for a ratio of 184. These extremes are influenced by differences in racial composition and by average levels of income, health care, morbidity, and other conditions that affect mortality by race. For example, in the nation's capital nearly 70 percent of the population is black and over a fifth of that group is below the poverty line. In Hawaii, on the other hand, almost a third of the population is Japanese and Chinese, and less than a twentieth of those groups is poor.

Socioeconomic variations associated with racial composition are not the only influence on mortality, however, for while North Dakota has the lowest age-adjusted death rate for females and the second lowest for males and while only 5 percent of its population consists of other races than white, the poverty rate is still about 25 percent higher than the national average for whites. Many other factors influence mortality, such as the proportions that are rural, urban, and suburban; the quality of water, air, and other resources; the economic base of the states and their counties and cities; and especially the quality and availability of local health care. Figure 2–4 also shows that one of the most striking features of state-to-state mortality variations is the roughly similar geographic pattern of male and female death rates. Thus, even though the range of age-adjusted death rates among males is considerably higher than among females, the two parts of Figure 2–4 resemble each other fairly closely: areas with high death rates for one sex generally have high rates for the other.

The highest death rates are in the states that extend from West Virginia and Kentucky into the Deep South, though Florida is an exception. The southern states still have comparatively high proportions of rural people, especially in small towns and villages with relatively high poverty rates and substandard health care. Even in many southern cities, however, effective death controls do not reach as many people as in some other parts of the nation. A second group of states with mortality rates above the national average for each sex extends westward from New York, New Jersey, and Delaware to other states with traditional centers of manufacturing—large urban agglomerations with significant inner-city poverty, relatively high rates of unemployment, and other conditions that adversely affect health and keep mortality rates high. Other states with relatively

Figure 2–4
Age-Adjusted Death Rate, by State and Sex, 1987

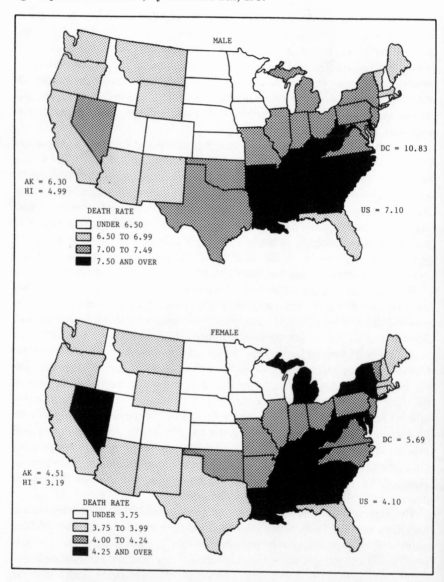

Source: U.S. Bureau of the Census, *Sixteenth Census of the United States: 1940, Characteristics of the Population, United States Summary* (Washington, DC: Government Printing Office, 1943), Table 7; Bureau of the Census, "State Population and Household Estimates, With Age, Sex, and Components of Change: 1981–88," *Current Population Reports*, Series P-25, No. 1044 (Washington, DC: Government Printing Office, 1989), Tables 7, 8; National Center for Health Statistics, *Vital Statistics of the United States, 1987*, Vol. II, *Mortality*, Part B (Washington, DC: Government Printing Office, 1989), Sec. 8, Table 6.

high death rates are Missouri and Oklahoma (and Texas for males). The lowest death rates for each sex appear in several states of the Midwest, where rural people are mostly middle class and many cities are comparatively small and less pressed by problems that plague huge metropolises elsewhere. Utah, Colorado, and Idaho also have very low death rates for each sex. Finally, relatively low rates occur in the Pacific Coast states, those of the extreme Southwest, and Montana and Wyoming—roughly the western third of the nation, though Nevada remains an island of relatively high mortality.

The sex mortality differential shows up in the states as it does in the regions and divisions. Thus, the age-adjusted sex mortality ratio is 140 in Alaska, 156 in Hawaii, and 159 in Maryland; the ratio is 160–175 in 25 other states; it is more than 175 in the remaining 22 and the District of Columbia (190). In every state and at every specific age the death rate of males is significantly higher than that of females, though the difference is smallest for very young and very old people and largest for adolescents and young adults. This pattern in the states repeats the pattern in the divisions, the regions, and the nation as a whole. This pattern is especially remarkable, given the higher poverty rates of females. In 1987, 15 percent of all females were poor, compared with 12 percent of all males; among female-headed families with no husband present the poverty rate was 34 percent, contrasted with just 6 percent in married-couple families.[78] Thus, females have lower death rates than males despite women's relative economic disadvantages and accompanying health risks, and if those problems were solved the sex mortality ratio would be even greater than it is.

With a few exceptions, the lowest male/female mortality ratios do not appear consistently in states with low age-adjusted death rates, and the highest ratios do not show up consistently in states with high death rates. At the extremes of the death rate there is a close relationship with the sex mortality ratio: Alaska and Hawaii, for example, have low rates and ratios, while the District of Columbia has the highest of both. Otherwise, the many conditions that affect both death rates and sex mortality ratios do not create a consistent pattern by state.

TRENDS

Several important trends in general mortality are (1) a precipitous decline during the twentieth century; (2) plateaus, steep drops, and periodic increases during that long decline; (3) faster decreases among females than males, at least until the 1980s; (4) a rather ragged but inexorable long-term tendency for the death rate of blacks to become more like that of whites, though with regress among black males after 1960; and (5) differential fluctuations in the death rates of various age groups because of aging of the population, though not in ways often assumed.

As noted in Chapter 1, age-adjusted death rates can be computed only from 1900 to the present because "the annual collection of mortality statistics for the national death-registration area began with the calendar year 1900."[79] Furthermore, some year-to-year mortality fluctuations prior to 1933 and even later are due to imperfect registration and compilation, redefinition of the race categories, the admission of additional states to the registration system, and other methodological problems. The separation of blacks from other races is especially important. Until the early 1960s virtually all mortality data that specified race were provided for "whites" and "other races." Many still are. For much of the period covered by mortality trends blacks were 95 percent or so of all nonwhite groups, so "other races" essentially meant "black," but Native Americans, Japanese, Chinese, and other racial groups have grown and their variable death rates make it more difficult to draw conclusions about the overall group of other races. Consequently, since the early 1960s blacks have been identified separately by the NCHS in some data reporting, but even in the late 1980s only about half the mortality data for other races were separated for blacks and others.[80] In the following sections, except the last, consistency over time requires use of the other-races category, and while that group is largely blacks and the comparisons are stated as applying to blacks and whites, one should keep in mind the diverse mortality patterns of other races, especially in the 1980s, when assessing trends. This problem aside, the trends do reflect some spectacular changes in the twentieth century.

Long-Term Reduction

Between 1900 and 1988 the age-adjusted death rate of the entire population, undifferentiated by sex or race, fell from 17.8 to 5.4, or 70 percent, which represents movement through the epidemiologic transition discussed in Chapter 1. The rate of decrease differed for various groups, but all experienced substantial reductions (see Figure 2–5). These groups also went through some periods of leveling and even increases in about the same way, so four general stages typify their twentieth-century experience.

1. The death rate fell fairly consistently from 1900 to about 1954, though with a pronounced rise in 1917–18 because of the influenza epidemic. There were earlier and later fluctuations too, but none that compromised for long the overall decline during the first half of the century—a decline that took place largely because of improvements in sanitation, nutrition, and the quality and availability of medical care. Life expectancy at birth increased significantly between 1900 and 1954 for both sexes and both major races, although blacks gained more rapidly than whites, largely because blacks were at a much greater initial disadvantage and had more room to improve. Black females experienced particularly dramatic reductions in mortality, and by 1955 their age-adjusted death rate was the same as that for white males.

2. Between 1954 and the late 1960s the age-adjusted death rate barely changed for whites and for black males, though it did continue to fall somewhat for black females, thus pushing their rate well below that of white males, where it has remained. The rate for black males also fell during part of the period but then

Figure 2–5
Age-Adjusted Death Rate, by Race and Sex, 1900–88

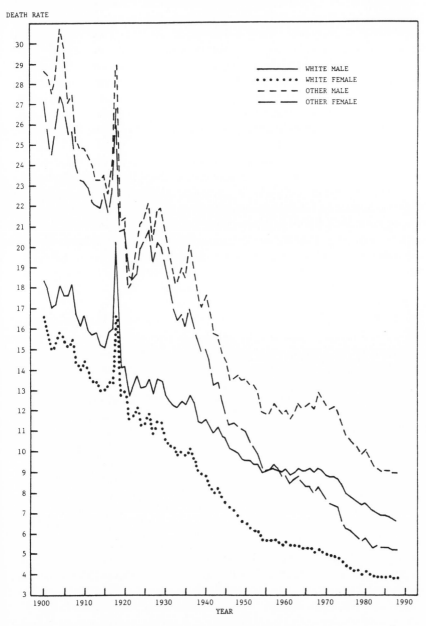

Source: National Center for Health Statistics, *Vital Statistics of the United States, 1987*, Vol. II, *Mortality*, Part A (Washington, DC: Government Printing Office, 1990), Sec. 1, Table 2; NCHS, "Advance Report of Final Mortality Statistics, 1988," *Monthly Vital Statistics Report* 39:7 (Washington, DC: Government Printing Office, November 1990), Table 12.

rose again to a 1968 peak identical to the one in 1953. White males were no better off in 1968 than in 1955, though the rates for white females in the late 1960s were a little lower than at any prior time. The net result was some racial convergence in death rates, some divergence by sex, but limited or no progress for the major groups.

3. After 1968 the death rate again fell sharply, especially for black males and females but also white males, while reductions for white females were least pronounced. As a result, the death rate of black females continued to drop below those of black and white males and toward the rate of white females. There was some convergence in the 1970s between black and white males as well, but less than among females. Part of the improvement during this period resulted from new social programs in the 1960s. Medicare, instituted to reduce the personal financial burden of health care for the elderly, helped lower their death rate, and by 1980 over 95 percent of all elderly persons both above and below the poverty line were covered by Medicare and over 99 percent had Medicare, private medical insurance, or both. Other programs and overall improvements in the economy helped other groups benefit from medical technology, and their death rates fell. The poverty rate of persons also dropped from 17.3 percent in 1965 to 11.7 percent in 1979, though whites had a greater decrease than blacks. No matter what the forces, the death rates of all groups decreased significantly between 1968 and 1979.

4. In 1979 the death rates of all groups rose somewhat, fell measurably for two years, and then tended to level out. The progress after 1982 was somewhat greater for white males than for the other three groups, but it was modest for each of them. The early 1980s were a time of recession, and all of the 1980s witnessed decreased funding for some social programs, funding for others at rates below inflation, and more exclusionary eligibility requirements. These and other forces, such as health damage from drugs, alcohol, cigarettes, and HIV/ AIDS, limited further reductions in death rates and even produced some increases.

Changing Race and Sex Differentials

Despite long-term decreases in death rates of males and females and blacks and whites, the pace of decrease varied and mortality differentials changed. Figure 2–6 shows the principal changes, using black/white mortality ratios by sex and male/female mortality ratios by race.

Death rates among females of both races fell much more than those among males; the sex mortality ratio among whites increased from 110 in 1900 to a high of 185 in 1979, and that among blacks rose from 106 in 1900 to a peak of 180 in 1981. There were some reductions, however, as the female mortality advantage fell temporarily, most notably between 1919 and 1921, in 1939, and again in the early 1980s, largely for reasons discussed earlier. In fact, in 1919–21 the male/female mortality ratio for blacks even dropped below 100 because

Figure 2–6
Male/Female and Black/White Mortality Ratios, 1900–88

Source: National Center for Health Statistics, *Vital Statistics of the United States, 1987*, Vol. II, *Mortality*, Part A (Washington, DC: Government Printing Office, 1990), Sec. 1, Table 2; NCHS, "Advance Report of Final Mortality Statistics, 1988," *Monthly Vital Statistics Report* 39:7 (Washington, DC: Government Printing Office, November 1990), Table 12.

briefly the death rate was lower for males than females. Furthermore, the sex mortality ratios of blacks and whites were at almost the same point in 1900 because women of both races shared about the same advantage relative to men in their respective races. Then, however, the mortality situation of white women relative to white men improved faster than that of black women relative to black men, so the ratios diverged significantly until the late 1960s. Black women then began to "catch up," and by 1988 the sex mortality ratios of the races were nearly the same again, though at a much higher figure than in 1900. In 1900 life expectancy at birth of males and females differed by only 2.9 years among whites and 2.5 years among blacks, whereas by 1980 the differences were 7.4 years and 8.8 years, respectively; by 1988, as mentioned earlier, the differences had fallen to 6.6 years for whites and 8.5 years for blacks. Nonetheless, the most striking long-term trend in sex mortality ratios is the significant increase for both races.

The mortality ratios by race fluctuated more than those by sex (see Figure 2–6). Just as the rising male/female mortality ratio for both races reveals a better mortality situation for women than for men, so the gradually declining black/white mortality ratio for both sexes shows an improving mortality situation among blacks relative to whites. That trend represents progress, but the improvement was not as rapid as it might have been and it was punctuated by many reversals, some lasting several years.

In 1900 the black/white mortality ratio was 161 among females and 156 among males. By 1988 those discrepancies had declined to 137 and 134, respectively. At times during the century, the position of blacks relative to whites either deteriorated so much or failed to improve that the racial mortality ratio reached 181 among females in 1930 and 171 among males in 1904. Some of the early increases, however, were because several southern states were added to the death-registration area, and the high death rates of their large, poor black populations worsened the statistical situation of the nation's whole black population. In 1925, for example, Alabama was admitted to the death-registration system and the national black/white mortality ratio for both sexes rose sharply. The opposite effect occurred in 1925 when Georgia was dropped from the system (it was readmitted in 1928).[81] After 1933 that factor no longer affected the mortality data.

Black females fared much worse in their comparison with white females than black males fared in their comparison with white males. Consequently, at no time during the century has the black/white mortality ratio among females fallen below the black/white ratio among males, and the two figures never even came close until the 1970s and 1980s. White females consistently have had lower age-adjusted death rates than any other race and sex group and their favorable mortality situation proved impossible for black females to duplicate, given the poorer death controls available to blacks as a whole. This is reflected in life expectancy at birth. In 1900 white females could expect to live 16.0 years longer than black females and even as late as 1968 the discrepancy was 7.2 years. The difference for males was 15.7 years in 1900 and 7.9 years in 1968. This situation changed after the late 1960s, however, and the black/white mortality ratio for females fell appreciably, even while the black/white ratio for males remained about the same. By 1988 the life expectancy differential between black and white females had shrunk, while that between black and white males had grown. Thus, the long-term disadvantage of black women relative to white women finally diminished, though a significant differential remains.

If black males are the focus of sex mortality differentials, it is clear they still fare worse than the other three groups, just as they did in all years except 1919–21. They are also losing ground relative to the other groups. In 1960, for example, the black/white mortality ratio among males was 130, but by 1988 it was 138. The mortality ratio between white females and black males increased from 216 in 1960 to 236 in 1988, while that between black males and black females grew from 136 to 173. Thus, while all four groups experienced significant death rate reductions after 1960, those for black males were the smallest and their relative mortality situation deteriorated. The reasons are basically socioeconomic, and while genetic factors may be partly responsible (e.g., sickle-cell anemia), those influences are small compared with differences in death rates for homicide, HIV/AIDS, and other preventable hazards. The biological differences by race are not nearly as significant as the biological differences that separate death rates of males and females.

Changing Age Differentials

The age-adjusted death rate does not show the different rates at which mortality has changed in various age categories. Age-specific death rates do, however, so Table 2–7 presents them by race and sex in 1970–72 and 1985–87. The period 1970–72 was chosen because the reduction in overall death rates resumed then after almost 15 static years and because the data are available for blacks separate from other races. Three-year averages reduce the effect of annual fluctuations in mortality and the possibility of significant misreporting in any one year.

The death rate decreased faster for infants than any other age group, regardless of sex or race, except for black females aged 35–44. More broadly, declines in the death rate were greater for children than other age groups; the deaths children do incur are largely from violence and accidents, especially motor vehicle accidents.[82] The rates of decrease were almost the same for white male and female infants, as were those for black male and female infants, though in both races the death rates themselves remain higher for males than females. Judging from the similar reductions, while the biological causes of greater longevity among females operate even in the first hours and days of life, the effect of controlling major causes of death among persons aged 0–14 is relatively sex neutral.

Among white males the pace of decrease in death rates was very slow in the

Table 2-7

Change in Age-Specific Death Rate, by Race and Sex, 1970-72 to 1985-87

Race and Age	Male[a]			Female[a]		
	1970-72	1985-87	Percent Change	1970-72	1985-87	Percent Change
White						
0-1	1,923.5	985.9	-48.7	1,455.3	763.0	-47.6
1-4	82.4	52.2	-36.7	65.2	40.3	-38.2
5-14	47.0	29.9	-36.4	30.4	18.6	-38.8
15-24	168.9	139.8	-17.2	60.8	49.5	-18.9
25-34	173.8	164.6	-5.3	82.4	60.5	-26.6
35-44	334.4	246.5	-26.3	190.2	120.6	-36.6
45-54	868.4	594.6	-31.6	452.4	331.8	-26.7
55-64	2,166.0	1,580.1	-27.0	1,002.8	858.1	-14.4
65-74	4,747.0	3,633.3	-23.5	2,421.3	2,020.6	-16.5
75-84	10,037.6	8,351.4	-16.8	6,558.5	5,117.8	-22.0
85 and over	19,804.0	18,600.0	-6.1	16,482.3	14,522.9	-11.9
Black						
0-1	3,884.1	2,176.0	-44.0	3,124.2	1,759.7	-43.7
1-4	142.7	90.1	-36.9	119.4	73.6	-38.4
5-14	65.0	41.9	-35.5	42.6	26.7	-37.3
15-24	316.9	189.5	-40.2	110.5	63.9	-42.2
25-34	555.0	374.3	-32.6	226.5	144.3	-36.3
35-44	940.5	673.1	-28.4	513.4	288.2	-43.9
45-54	1,749.9	1,271.1	-27.4	1,015.3	651.6	-35.8
55-64	3,222.1	2,544.4	-21.0	1,921.3	1,472.2	-23.4
65-74	5,778.8	4,805.4	-16.8	3,831.4	2,897.5	-24.4
75-84	8,985.4	9,276.6	+3.2	6,237.5	6,182.2	-0.9
85 and over	13,923.3	15,253.5	+9.6	11,574.3	12,326.1	+6.5

Source: U.S. Bureau of the Census, "United States Population Estimates, by Age, Sex, Race, and Hispanic Origin, 1980 to 1988," Current Population Reports, Series P-25, No. 1045 (Washington, DC: Government Printing Office, 1990), Table 1; National Center for Health Statistics, Vital Statistics of the United States, 1987, Vol. II, Mortality, Part A (Washington, DC: Government Printing Office, 1990), Sec. 1, Table 4.

[a]Deaths per 100,000 population in specified group.

ages 25-34, largely because of high losses to major vehicle accidents; the pace of decrease rose for men in their 30s and 40s, though not to the point reached in infancy and childhood. The pace of reduction then remained fairly constant until age 65 or so, after which it gradually dropped off to very modest change in the ages 85 and over. These patterns reflect higher success rates in preventing and treating heart disease, stroke, and other causes of death among middle-aged and young-old persons than among the very old. Part of the leveling that shows up after age 85 is also due to the increasing percentage of persons 90, 95, and even 100, whose death rates are very high and for whom age reporting is often inaccurate.

Among white females the pace of mortality reduction with age was roughly similar to that among white males. Females also experienced lower rates of improvement in their late teens and early 20s and in the ages 75 and over than they did during infancy and childhood, but fluctuations with age were generally less abrupt than for males. In addition, the pace of reduction in death rates among women aged 45–64 was relatively slow and that lag helped narrow the sex differential in life expectancy. Those are the ages when the death rate for lung cancer has risen among women.

Among black males and females the pace of mortality reduction was greatest in infancy, childhood, and young adulthood, followed by loss of momentum up to age 75 and by increases in the death rate of men 75 and over and women 85 and over. Part of the increase was due to rising proportions of persons 90, 95, and older and to age misreporting, also experienced by whites, but part was due to deterioration in socioeconomic status of the black old-old compared with the white old-old. Those conditions reduced the magnitude of the mortality crossover and may even eliminate it. In 1970–72, for example, the death rate of black males aged 85 and over was 30 percent less than that of white males, but by 1985–87 it was only 18 percent less; among females the gap decreased from 30 to 15 percent. Thus, the crossover still occurs, but blacks have a smaller reported advantage now than in the past.

Since the pace of mortality reduction among elderly people was less than among infants, children, and even middle-aged adults, falling mortality among the elderly cannot be the primary factor in aging the population. Yet ours is an aging population in that the percentage of persons 65 and over is growing and the median age is rising. In 1900 the elderly were 4 percent of the population; in 1990 they were nearly 13 percent; by 2050 they are projected to become 23 percent. The median age rose from 22.9 years in 1900 to 33.0 years in 1990; it is expected to reach 42.7 years by 2050 and to climb slightly for several decades thereafter.[83]

This aging process is largely attributable to changes in fertility, though mortality reductions do play a secondary role. The total fertility rate (average number of children per woman) fell in the 1970s and earlier large cohorts grew older, but they were not replaced by proportionately large cohorts of newborns. Consequently, since a total population can only equal 100 percent, children became a smaller percentage and other age groups larger percentages. Eventually, the baby boom population born between 1945 and the early 1960s will enter the age group 65 and over and it will grow rapidly as a proportion of the total; so will the very old.

The role played by decreasing mortality among the elderly, even at relatively modest rates, does keep more of them alive longer and increases their number. So do increases in life expectancy at birth and other ages. Death now generally occurs at older ages than in the past: by 1988, 71 percent of all deaths occurred in the age group 65 and over (21 percent in the ages 85 and over), but as recently as 1940 only 46 percent of all deaths happened to people 65 and over and a mere 6 percent struck those 85 and over. Despite the progress of the past several decades, however, elderly people have disproportionately high levels of chronic illness, so life-sustaining treatments have a poorer outcome for them than for younger people. When a life-threatening condition strikes an elderly person, especially in the old-old ages, the combination of that condition with one or more chronic illnesses complicates treatment. Across the entire elderly population, this complex interplay of factors holds down the pace of mortality reduction.

Increases in chronological age are coupled with reductions in physiological reserve—efficiency and functioning of the heart, lungs, kidneys, liver, and other major organs. They are more subject to declines in such functions as cardiac index, renal plasma flow, maximal breathing capacity, and basal metabolic rate. Those declines are not illnesses in themselves, but they do decrease the body's ability to rebound from acute illness or trauma and thereby reduce the prospects for a favorable treatment outcome.[84] These conditions make the pace of reduction in mortality less for the elderly than for the young, so the reductions that do occur among the elderly are not responsible for their growing proportion. That would only be so if the death rates of elderly people were falling faster than those of younger people. In fact, the aging of the population goes on despite differential mortality reduction by age, not because of it.

NOTES

1. U.S. Bureau of the Census, Ellen Jamison, Peter D. Johnson, and Richard A. Engels, *World Population Profile: 1987* (Washington, DC: Government Printing Office, 1987), p. 10.

2. Ibid.

3. U.S. Bureau of the Census, *Social Indicators III* (Washington, DC: Government Printing Office, 1980), p. 37.

4. For the data, see U.S. Bureau of the Census, Ellen Jamison, *World Population Profile: 1989* (Washington, DC: Government Printing Office, 1989), p. 33.

5. John Paxton, ed., *The Statesman's Yearbook, 1988–89* (New York: St. Martin's Press, 1988), p. 387.

6. Paul B. Goodwin, Jr., ed., *Latin America*, 3d ed. (Guilford, CT: Dushkin Publishing Group, 1988), p. 25.

7. Louis J. Dublin, Alfred J. Lotka, and Mortimer Spiegelman, *Length of Life* (New York: Ronald Press, 1949), pp. 129–30.

8. Francis C. Madigan, "Are Sex Mortality Differentials Biologically Caused?" *Milbank Memorial Fund Quarterly* 35:2 (April 1957), pp. 129–130.

9. George J. Stolnitz, "A Century of International Mortality Trends: II," *Population Studies* 10:1 (March 1956), pp. 22–32.

10. John R. Weeks, "The Demography of Islamic Nations," *Population Bulletin* 43:4 (Washington, DC: Population Reference Bureau, 1988), pp. 28, 30–31.

11. Robert D. Retherford, *The Changing Sex Differential in Mortality* (Westport, CT: Greenwood Press, 1975), p. 9.

12. Samuel H. Preston, *Mortality Patterns in National Populations With Special Reference to Recorded Causes of Death* (New York: Academic Press, 1976), p. 121.

13. Ibid., p. 125.

14. Retherford, *Changing Sex Differential*, pp. 12–13.

15. Robert D. Retherford, "Tobacco Smoking and the Sex Mortality Differential," *Demography* 9:2 (May 1972), pp. 203–216. See also Retherford, *Changing Sex Differential*, Chap. 6.

16. For the data, see National Center for Health Statistics, *Health, United States, 1987* (Washington, DC: Government Printing Office, 1988), Table 26; and NCHS, "Advance Report of Final Mortality Statistics, 1988," *Monthly Vital Statistics Report* 39:7 (Washington, DC: Government Printing Office, November 1990), Table 12.

17. For a discussion of the decreasing sex difference in life expectancy, see Metropolitan Life Insurance Company, "Women's Longevity Advantage Declines," *Statistical Bulletin* 69:1 (January-March 1988), pp. 18–23.

18. Rita Rubin, "The Ways of a Woman's Heart," *American Health* 10:2 (March 1991), p. 6.

19. National Center for Health Statistics, Charlotte A. Schoenborn and Bernice H. Cohen, "Trends in Smoking, Alcohol Consumption, and Other Health Practices Among U.S. Adults, 1977 and 1983," *Advance Data from Vital and Health Statistics* 118 (Washington, DC: Government Printing Office, June 1986), p. 8.

20. National Center for Health Statistics, *Health, United States, 1989* (Washington, DC: Government Printing Office, 1989), p. 8.

21. Ibid.

22. Robert A. Weinberg, "Oncogenes and Tumor Suppressor Genes," in *Unnatural Causes: The Three Leading Killer Diseases in America*, edited by Russell C. Maulitz (New Brunswick, NJ: Rutgers University Press, 1989), p. 83.

23. Bryan L. Boulier and Vincente B. Paqueo, "On the Theory and Measurement of the Determinants of Mortality," *Demography* 25:2 (May 1988), p. 251.

24. Lois M. Verbrugge, "A Health Profile of Older Women with Comparisons to Older Men," *Research on Aging* 6:3 (September 1984), p. 291.

25. National Center for Health Statistics, "Current Estimates From the National Health Interview Survey: United States, 1987," *Vital and Health Statistics*, Series 10, No. 166 (Washington, DC: Government Printing Office, 1988), Table 2.

26. Ibid., Table 58.

27. Tom Hickey, William Rakowski, and Mara Julius, "Preventive Health Practices Among Older Men and Women," *Research on Aging* 10:3 (September 1988), p. 316. See also Eileen M. Crimmins, Yasuhiko Saito, and Dominique Ingegneri, "Changes in Life Expectancy and Disability-Free Life Expectancy in the United States," *Population and Development Review* 15:2 (June 1989), pp. 247–248.

28. For the data, see NCHS, "Current Estimates," Tables 17, 27, 37, 67.

29. Hickey et al., "Preventive Health Practices," p. 316.

30. Ibid.

31. Crimmins, et al., "Changes in Life Expectancy," p. 248.

32. Kenneth G. Manton and Eric Stallard, *Recent Trends in Mortality Analysis* (New York: Academic Press, 1984), p. 155.

33. U.S. Bureau of the Census, "Demographic Aspects of Aging and the Older

Population in the United States," *Current Population Reports*, Series P–23, No. 59 (Washington, DC: Government Printing Office, 1978), p. 32.

34. Ansley J. Coale and Graziella Caselli, "Estimation of the Number of Persons at Advanced Ages from the Number of Deaths at Each Age in the Given Year and Adjacent Years," *Genus* 46:1 (January-June 1990), p. 1.

35. U.S. Bureau of the Census, "Demographic Aspects of Aging," p. 32. See also Charles B. Nam, Norman L. Weatherby, and Kathleen A. Ockay, "Causes of Death Which Contribute to the Mortality Crossover Effect," *Social Biology* 25:4 (Winter 1978), pp. 306–314.

36. Adapted from Manton and Stallard, *Mortality Analysis*, p. 246.

37. Ibid., p. 278.

38. Jacquelyn Johnson Jackson, *Minorities and Aging* (Belmont, CA: Wadsworth, 1980), p. 64.

39. U.S. Bureau of the Census, Arlene F. Saluter, "Marital Status and Living Arrangements: March 1988," *Current Population Reports*, Series P–20, No. 433 (Washington, DC: Government Printing Office, 1989), Table 6.

40. Lisa Berkman, Burton Singer, and Kenneth Manton, "Black/White Differences in Health Status and Mortality Among the Elderly," *Demography* 26:4 (November 1989), pp. 674–676.

41. Ibid., p. 676.

42. Ibid., pp. 662–663.

43. For an analysis, see John Reid, "Black America in the 1980s," *Population Bulletin* 37:4 (Washington, DC: Population Reference Bureau, 1982), p. 15.

44. For a study of this matter, see Eugene Schwartz, Vincent Y. Kofie, Marc Rivo, and Reed V. Tucson, "Black/White Comparisons of Deaths Preventable by Medical Intervention: United States and the District of Columbia," *International Journal of Epidemiology* 19:3 (September 1990), pp. 591–598.

45. U.S. Bureau of the Census, *1980 Census of Population, Race of the Population by States: 1980*, Supplementary Reports, PC–80–S1–3 (Washington, DC: Government Printing Office, 1981), Table 1.

46. National Center for Health Statistics, *Health, United States, 1979* (Washington, DC: Government Printing Office, 1980), p. 10.

47. Elena S. H. Yu, "Health of the Chinese Elderly in America," *Research on Aging* 8:1 (March 1986), pp. 84, 94. See also Charles B. Nam and Susan Gustavus Philliber, *Population: A Basic Orientation*, 2d ed. (Englewood Cliffs, NJ: Prentice-Hall, 1984), pp. 86–87.

48. Yu, "Health of the Chinese Elderly," p. 95.

49. Ibid., p. 96.

50. Ibid., p. 84.

51. U.S. Bureau of the Census, "Hispanic Population of the United States: March 1989," *Current Population Reports*, Series P–20, No. 444 (Washington, DC: Government Printing Office, 1990), p. 30.

52. NCHS, "Advance Report of Final Mortality Statistics, 1988," p. 47.

53. Ira Rosenwaike, "Mortality Differentials among Persons Born in Cuba, Mexico, and Puerto Rico Residing in the United States," *American Journal of Public Health* 77:5 (May 1987), pp. 603–606.

54. Ira Rosenwaike and Katherine Hempstead, "Mortality Among Three Puerto Rican Populations: Residents of Puerto Rico and Migrants in New York City and the Balance

of the United States, 1979–81,'' *International Migration Review* 24:4 (Winter 1990), p. 694.

55. Ibid., pp. 685–687. Another study of racial and ethnic differentials is Joel C. Kleinman, ''Infant Mortality Among Racial/Ethnic Minority Groups, 1983–1984,'' *Morbidity and Mortality Weekly Report* 39:553 (July 1990), pp. 31–39.

56. NCHS, ''Current Estimates,'' p. 9.

57. Walter R. Gove, ''Sex, Marital Status, and Mortality,'' *American Journal of Sociology* 79:1 (July 1973), pp. 45, 65.

58. Yuanreng Hu and Noreen Goldman, ''Mortality Differentials by Marital Status: An International Comparison,'' *Demography* 27:2 (May 1990), p. 246.

59. John R. Weeks, *Population: An Introduction to Concepts and Issues*, 4th ed. (Belmont, CA: Wadsworth, 1989), p. 167.

60. Gove, ''Sex, Marital Status, and Mortality,'' p. 46. See also Hu and Goldman, ''Mortality Differentials by Marital Status,'' p. 246.

61. Gove, ''Sex, Marital Status, and Mortality,'' pp. 45, 65.

62. Ibid., pp. 49–54.

63. Ibid., pp. 54–56.

64. Ibid., pp. 56–59. For more recent data on the number of deaths in each marital category by cause, age, gender, and race, see National Center for Health Statistics, *Vital Statistics of the United States, 1987*, Vol. II, *Mortality*, Part A (Washington, DC: Government Printing Office, 1990), Sec. 1 Table 32. For the population estimates necessary to compute age-specific death rates, see Bureau of the Census, ''Marital Status: 1987,'' Table 1.

65. NCHS, ''Current Estimates, p. 9.''

66. Ibid.

67. Kitagawa and Hauser, *Differential Mortality*, pp. 23–24.

68. U.S. Bureau of the Census, Paul C. Glick, ''Population of the United States, Trends and Prospects: 1950–1990,'' *Current Population Reports*, Series P–23, No. 49 (Washington, DC: Government Printing Office, 1974), pp. 46–47, based on data in Kitagawa and Hauser, *Differential Mortality*.

69. Harriet Orcutt Duleep, ''Measuring Socioeconomic Mortality Differentials Over Time,'' *Demography* 26:2 (May 1989), p. 345.

70. Ibid., p. 347.

71. For the data, see U.S. Bureau of the Census, ''Educational Attainment in the United States: March 1987 and 1986,'' *Current Population Reports*, Series P–20, No. 428 (Washington, DC: Government Printing Office, 1988), Table 1.

72. Ibid.

73. Kitagawa and Hauser, *Differential Mortality*, pp. 12, 18.

74. National Center for Health Statistics, *Health, United States, 1981* (Washington, DC: Government Printing Office, 1981), p. 210.

75. Henry S. Shryock and Jacob S. Siegel, *The Methods and Materials of Demography*, Vol. 2 (Washington, DC: Government Printing Office, 1973), pp. 409–410.

76. See Vincente Navarro, ''Race or Class Versus Race and Class: Mortality Differentials in the United States,'' *Lancet* 336:8725 (November 17, 1990), pp. 1238–1240.

77. For an account of the geographic distribution of poverty, see Paul E. Zopf, Jr., *American Women in Poverty* (Westport, CT: Greenwood Press, 1989), pp. 47–58.

78. U.S. Bureau of the Census, ''Poverty in the United States: 1987,'' *Current Pop-

ulation Reports, Series P–60, No. 163 (Washington, DC: Government Printing Office, 1989), Tables 7, 13.

79. Robert D. Grove and Alice M. Hetzel, *Vital Statistics Rates in the United States, 1940–1960* (Washington, DC: Government Printing Office, 1968), p. 7.

80. NCHS, *Mortality*, Sec. 7, p. 6.

81. Ibid., p. 9.

82. National Center for Health Statistics, Lois Fingerhut, "Trends and Current Status in Childhood Mortality, United States, 1980–85," *Vital and Health Statistics*, Series 3, No. 26 (Washington, DC: Government Printing Office, 1989), p. 31.

83. For the data, see U.S. Bureau of the Census, Gregory Spencer, "Projections of the Population of the United States, by Age, Sex, and Race: 1988 to 2080," *Current Population Reports*, Series P–25, No. 1018 (Washington, DC: Government Printing Office, 1989), Table 4; and U.S. Congress, Office of Technology Assessment, *Life-Sustaining Technologies and the Elderly* (Washington, DC: Government Printing Office, 1987), p. 77.

84. Ibid., pp. 78–79.

3 Infant Mortality

Infant mortality reflects the well-being of whole populations, whether nations or subgroups, which makes it a fundamental area of mortality study. The level of infant mortality also is closely associated with a population's fertility: where infant mortality rates are high, fertility rates also often are high. One factor in such cases is the "replacement effect," in which parents still in the reproductive ages produce additional children to replace those who die. Another factor is the "insurance effect," in which parents do not begin spacing or limiting births until they feel they have enough children to reach adulthood and provide the elders with "social security." Such groups often favor early marriage and high rates of reproduction in response to high rates of infant and child mortality. This is especially common where the rural population is proportionately large and average levels of living are low, since both conditions create a perceived need for many births. Thus, the infant mortality rate reflects levels of living and has a causal effect on fertility.

Infant mortality rates have declined sharply in most developing countries in the past several decades, just as they did earlier in the developed countries. The reductions resulted from improvements in sanitation, nutrition, medical technology, and health care, such as prevention and treatment of diarrhea, which is a major killer of infants in poor socioeconomic conditions. As more efficient death controls expand in a nation, the newborn are most affected, so the percentage of infants and children grows, along with the average support burden that productive adults must carry. This situation is winding down in some developing nations, however, as birth rates fall and children decline as a percentage of the total population, while other age groups, especially the elderly, become larger percentages. Thus, the study of infant mortality globally and in the United

States reveals significant reductions that affect population growth rates and age profiles; it also shows persistently high rates in some groups.

SOME INTERNATIONAL COMPARISONS

In 1988–90 the world infant mortality rate was 75 deaths per 1,000 live births— a high figure relative to those in places with sophisticated prenatal and postnatal care and high average levels of living.[3] In fact, while the world is demographically diverse, much of it is divided into two segments identified by differential infant mortality rates and their pace of decline. One segment includes countries with little or no population growth and high average levels of living; the other segment comprises nations with rapidly growing populations and low average levels of living. The first group of countries has low infant mortality rates; the second group has high rates that may even rise if the relationship between population and ecological support systems deteriorates and lowers the level of living for large numbers of people. Countries actually range along a continuum from most developed to least developed, but the bifurcation is real enough. Thus, in 1988– 90 infant mortality rates were 93 in the developing countries (excluding China), 42 in China, and only 15 in the developed countries.[5] Furthermore, progress in the 1980s was faster in the developed countries, where the infant mortality rate dropped almost twice as fast as in the developing ones, although it fell almost everywhere during the past few decades.

The infant mortality gap between developed and developing countries is illustrated in Table 3–1. It shows 15 countries with very low infant mortality rates in 1988–90 and 15 with high rates, along with the annual percent natural increase, total fertility rate (TFR), and per capita Gross National Product (GNP).[6] GNP is an imperfect index of the level of living because it does not reflect class differences within a country, value of subsistence commodities, advantages derived from the "informal economy," or benefits sometimes provided by governments. GNP also excludes the food-producing and other unrecorded economic activities of a large share of the world's women.

Per capita GNP is one basis for comparing the average level of affluence in countries with low infant mortality rates and those with high rates, but it is only one of many factors that affect mortality. No matter how income is measured, even if it were distributed more equitably among the world's nations or within a specific nation, other factors might sustain high infant mortality rates. They include the degree of freedom for women to make choices, the extent to which children of each sex are valued, the degree of emphasis on education, and the level of local activism and the ways it affects care for mothers and infants. For example, China and Benin have about the same per capita GNP, but the infant mortality rate in China is only a third that of Benin, so other factors than national income are at work. Consequently, infant mortality levels in many developing countries can be lowered substantially without major additional expenditures. Despite the disclaimer, however, very low per capita GNP often means that

Table 3–1
High and Low Infant Mortality Rates, Selected Countries, 1988–90

Country	Infant Mortality Rate	Percent Natural Increase	Total Fertility Rate[a]	Per Capita GNP 1988 ($US)
Highest rates				
Afghanistan	179	2.5	6.9	250
Sierra Leone	163	2.2	6.4	240
Mali	156	2.9	6.9	230
Gambia	153	2.4	6.4	220
Guinea	151	2.4	6.2	350
Central African Republic	146	2.5	5.7	390
North Yemen	145	3.4	7.9	650
Mozambique	145	2.6	6.3	100
Ethiopia	142	2.4	6.4	120
Angola	141	2.6	6.4	500[b]
Malawi	140	3.3	7.5	160
Chad	139	2.2	5.5	160
Niger	139	2.9	7.1	310
Somalia	139	2.9	7.1	170
Burkina Faso	138	2.9	6.7	230
Lowest rates				
Japan	5.0	0.5	1.7	21,040
Sweden	5.8	0.2	1.9	19,150
Finland	5.8	0.3	1.6	18,610
Switzerland	6.8	0.3	1.5	27,260
Hong Kong	7.6	0.8	1.4	9,230
Netherlands	7.6	0.4	1.6	14,530
France	7.7	0.4	1.8	16,080
Canada	7.7	0.7	1.7	16,760
Singapore	7.9	1.2	1.7	9,100
West Germany	8.1	-0.1	1.4	18,530
Denmark	8.2	0.0	1.5	18,470
Norway	8.2	0.2	1.8	20,020
Ireland	8.6	0.7	2.3	7,480
East Germany	8.7	0.0	1.7	10,400[b]
Luxembourg	8.7	0.1	1.4	22,600
United States	9.9	0.7	1.9	19,780

Source: Population Reference Bureau, World Population Data Sheet, 1988, 1989, 1990 (Washington, DC: Population Reference Bureau, 1988–90); John M. Paxton, ed., The Statesman's Yearbook, 1990–91 (New York: St. Martin's Press, 1990), pp. 63, 81.

[a]Average number of children born to a woman during her lifetime.

[b]For 1985.

many people have little or no access to rudimentary lifesaving measures, so their infant mortality rate is high.

The two groups of countries in Table 3–1 reflect tremendous divergence in infant mortality, natural increase, TFR, and per capita GNP at the poles of the demographically divided world. The infant mortality rate in 1988–90 ranged from 16 to 36 times higher in the 15 poorer countries than in the 15 wealthier ones, but 31 others also had infant mortality rates of 100 or more. In those 46 nations at least one of every ten newborns will die before the first birthday and many others will perish before they reach age five. Thirty-four countries with rates of 100 or more are in Africa, concentrated especially just below the Sahara Desert in a band stretching from the western coast to the Horn of Africa and into the Arabian Peninsula. Ten of the 46 countries are Asian and two (Haiti and Bolivia) are in Latin America, although that region as a whole has lowered its infant mortality rate dramatically since World War II.

Many countries with very high rates are Islamic nations; at least half the population is Muslim in eight of the developing countries in Table 3–1. In fact, the "Islamic nations as a group have the world's highest levels of mortality," despite their progress in saving lives since they became independent.[8] The reasons include high average levels of poverty, chronic underdevelopment, and a deficient resource base that retards efforts to lower mortality and raise the level of living.[9] These problems are partly reflected in per capita GNP, which is so meager in the 15 countries with high infant mortality that it is only 0.4 to 9 percent of per capita GNP in the 15 developed countries with low infant mortality.

The annual percent natural increase is much greater in the countries with high infant mortality rates than in those with low rates, and the TFR also differs dramatically. Several factors are involved.

1. The pace of natural increase is rapid in poor countries because infant mortality rates have already been reduced significantly—high though they may remain—thus allowing more infants to survive. In fact, since infant mortality rates in most developing countries have fallen far faster than birth rates, the pace of natural increase has accelerated. Guinea, for example, lowered its infant mortality rate from 216 in 1955 to 151 in 1988–90—a drop of 30 percent. During the same period the birth rate remained unchanged (47 per 1,000 population), so the decrease in infant mortality was the equivalent of a 30 percent rise in the birth rate and the rate of natural increase went up.

2. Places where infant mortality rates are still high have ample room for further reductions, though actual reductions will be affected by development, motivation and ability to use death controls, changes in the educational status of women, political conditions, and other factors. In some places potential decreases in infant mortality will become real, and unless birth rates decline at the same pace, the rate of natural increase will remain high. It could even increase, though recent trends suggest that it is more likely to persist or drop slightly over the short run and decrease significantly over the long run because of gradually falling fertility.

3. Current high rates of natural increase in many less developed countries do not necessarily cause underdevelopment, but high rates of natural increase result from various aspects of underdevelopment. Then, once underdevelopment has a strong grip on a large part of a country, population growth hinders efforts to develop.[10] Much, perhaps all, of a nation's annual economic gain may go to sustain more people at the same low level of living; poverty prevents infant mortality from falling as fast as it might; persistently high infant mortality encourages people to bear relatively large numbers of infants for their insurance effect. These conditions then create cultural lag, for even when infant mortality is falling, the rates of birth and natural increase may remain high.

In short, the infant mortality rate is intricately interwoven with other demographic and sociocultural conditions, and its level and pace of change affect and are affected by those conditions.

The same is true in highly developed countries, except that the interplay between demographic and sociocultural variables produces a low infant mortality rate, a very slow pace of natural increase, a TFR near or at replacement, and high per capita GNP. These conditions, however, produce lower infant mortality rates in 21 other countries than the United States. Sixteen are in Europe, but the list includes Japan, Hong Kong, Singapore, Canada, and Australia. Only one country (Switzerland) has a higher per capita GNP than the United States, so a high rank on that economic indicator does not guarantee excellent general health care for all segments of the population, nor full attention to low birth weight and other conditions that reduce an infant's survival potential. The United States ranks relatively low among the developed countries in reducing infant mortality because some segments of the population—most notably poor blacks—do not benefit equitably from death control measures or health insurance, and their relatively high infant mortality rates raise the national average.[11]

PATTERNS BY AGE OF VICTIM

The infant mortality rate is not distorted by the age profile of a population in the same way as the crude death rate, and for international comparisons the infant mortality rate not only serves well but is usually the only measure of infant deaths available. As an index of the infant death situation in the United States and of differences between various groups, however, the infant mortality rate is too unrefined to show the uneven distribution of infant deaths throughout the first year of life. Nor does it reveal anything about the loss of fetuses just prior to birth. For these purposes we can use indexes identified in Chapter 1: the perinatal mortality ratio, neonatal mortality rate, and postneonatal mortality rate. The first index, which bridges the birth trauma, combines deaths of infants 0–6 days with fetal deaths after 28 weeks gestation; the second refers to deaths under 28 days following birth; and the last accounts for deaths from 28 days to the end of the first year.

Table 3–2

Perinatal Mortality Ratio, by Residence, Race, and Sex, 1985–87

Residence and Sex	All Races[a]	White[a]	Black[a]
United States	10.4	9.1	17.2
Male	11.0	9.7	18.5
Female	9.7	8.6	15.9
Metropolitan	10.4	9.0	17.2
Nonmetropolitan	10.4	9.5	17.4

Source: National Center for Health Statistics, Vital Statistics of the United States, 1985, 1986, 1987, Vol. II, Mortality, Part A (Washington, DC: Government Printing Office, 1988-90), Sec. 4, Tables 2, 3.

[a]Perinatal mortality ratio is the number of deaths of infants under 7 days and of fetuses after 28 weeks gestation for each 1,000 live births in specified group (NCHS Definition I).

Perinatal Mortality Ratio

The perinatal mortality ratio mirrors the events of a particularly dangerous time for fetuses and newborns. In fact, about 53 percent of all fetal deaths occur after 28 weeks gestation and 54 percent of all infant deaths take place before the seventh day following birth. Therefore, the perinatal mortality ratio (10.4), shown in Table 3–2, happens to be identical to the infant mortality rate, which covers an entire year. Among whites the perinatal mortality ratio is slightly higher than the infant mortality rate and among blacks it is somewhat lower. This reflects two opposite tendencies. First, the loss of white fetuses is relatively concentrated late in pregnancy, whereas the loss of blacks is more evenly distributed throughout pregnancy. In 1985–87, for example, 56 percent of all fetal deaths among whites but only 46 percent among blacks occurred after 28 weeks gestation. Second, the deaths of black infants are somewhat more concentrated in the first few days after birth; 56 percent of all black infant deaths and 53 percent of all white infant deaths took place in the first six days. Thus, while black women are more likely to lose their fetuses in the early stages of pregnancy, the loss of black infants is somewhat more concentrated in their first few days, partly because a larger percentage is born with birth defects, AIDS, drug addiction, fetal alcohol syndrome, and other afflictions. These conditions reflect significant differences in the average socioeconomic environment of the races and the variable proportions that are poor. They also reflect differences among local areas in health care and various risks.

Several other factors are implicated in the perinatal mortality difference by race: frequency and quality of prenatal care, incidence of premature births, rate

Table 3-3
Percent of Live Births Under Specified Conditions, by Race of Child and Age of Mother, 1985–87

Race and Age	All Births	Prenatal Visits			Premature Birth[a]	Low Birth Weight[b]	Mother Unmarried
		None	1–6	Over 10			
White							
All ages	100.0	1.4	8.8	60.1	8.4	5.6	15.6
0–14	0.1	6.8	27.2	29.1	19.3	10.7	83.4
15–19	10.5	3.2	54.3	42.8	17.7	7.6	47.9
20–24	28.7	1.9	10.9	55.8	8.5	5.7	19.5
25–29	33.4	1.0	6.3	64.9	7.4	5.1	8.8
30–34	20.2	0.8	5.5	66.2	7.7	5.2	6.7
35–39	6.3	1.1	6.8	63.6	8.9	6.0	8.3
40–44	0.8	2.1	10.5	56.9	10.9	7.1	11.5
45–49	c	4.0	16.3	45.4	13.2	8.6	13.9
Black							
All ages	100.0	4.0	20.4	41.3	17.7	12.6	61.2
0–14	1.0	6.4	32.1	25.4	29.2	15.6	98.9
15–19	21.8	5.0	27.0	31.5	21.0	13.2	90.1
20–24	33.7	4.2	21.7	39.3	17.2	12.2	65.9
25–29	25.1	3.4	16.4	47.2	16.0	12.2	46.0
30–34	13.2	3.2	14.5	49.9	16.8	12.6	37.7
35–39	4.5	3.6	14.5	49.0	17.6	13.0	35.3
40–44	0.7	4.4	16.9	45.2	18.5	13.7	34.5
45–49	c	6.1	17.8	41.8	18.8	15.2	35.4

Source: National Center for Health Statistics, _Vital Statistics of the United States_, 1985, 1986, 1987, Vol. I, _Natality_ (Washington, DC: Government Printing Office, 1988–89), Sec. 1, Tables 39, 76, 89.

[a]Births occurring before 37 weeks gestation.

[b]Birth weight less than 2,500 grams.

[c]Less than 0.1 percent.

of low birth weight, and proportion of births to unmarried women.] Table 3–3 shows these factors, all of which reflect differences in the level of living and other components of socioeconomic status. Each factor refers to conditions under which live births occur; none involves spontaneous or induced abortions.

Prenatal Care. At most ages black women are more likely than whites to give birth without receiving professional prenatal care, although the racial difference is slight among mothers under 15. Even in the ages 30–34, when prenatal care is most likely for both races, only half the black women had more than ten visits to a health facility, compared with two-thirds of the whites. Furthermore, while prenatal care is more common among women of both races in their mid-20s and 30s than among very young women and even those in their 40s, a larger share of black than white births occurs toward the lower end of the age scale where prenatal care is least satisfactory. In 1985–87, for instance, teenagers produced 23 percent of all black births but only 11 percent of all white births. Therefore, even the age distribution of motherhood by race reflects higher risks for black than white fetuses and newborns. These risks help make the black perinatal mortality ratio nearly twice that of whites because illness rates and other conditions that contribute to perinatal mortality are much higher among teenage mothers than older women.[12] Moreover, a large share of adolescent pregnancies in both races is unintended, partly because many young women and men are ignorant about reproduction and birth control and harbor myths about pregnancy possibilities.[13] In turn, prenatal care and other protections that lower perinatal mortality are less adequate among unwanted teenage pregnancies carried to term than among planned pregnancies, especially those of postadolescent women. Even the young mothers themselves experience a higher incidence of complications and death in pregnancy and delivery than do older mothers.[14]

The timing of prenatal care also is crucial. Care during the first trimester of pregnancy reduces the risk of producing a low-birth-weight infant and thereby reduces its risk of death.[15] In 1987 about 21 percent of pregnant whites received no prenatal care during the first trimester, but almost 40 percent of blacks received none; the latter percentage was about the same for native Americans and Hispanics. Furthermore, the proportions decreased very little after 1978, and during the 1980s the incidence of early prenatal care scarcely changed for whites, blacks, native Americans, or Hispanics.[16]

Premature Births. Blacks are more likely than whites to produce "preterm" or "premature" births—those after fewer than 37 weeks gestation.[17] This is particularly true of very young mothers, and in 1985–87 about 29 percent of births to black mothers under 15 were premature, compared with 19 percent of births to whites, though the racial differential persists at all ages. The infants are apt to suffer from adverse conditions that caused prematurity in the first place, and those conditions, coupled with low birth weight and prematurity itself, help account for the high black perinatal mortality ratio. In the first 24 hours after birth, for example, the infant mortality rate among blacks is more than double that among whites, and while the discrepancy decreases in the following days, at all stages in the first six days black infants are at least one-third more likely than whites to succumb. Since premature birth is more common among very young mothers than older teenagers and women past 20, the heavier concentration of black childbearing in the younger ages also increases the incidence of prematurity and its risks. So do poor diet; tobacco, drug, and alcohol use during pregnancy; poor local health services; and other hazards.[18] These problems are most severe among the poor, and higher percentages of black than white teenagers are poor, so the survival prospects of black infants and the marital and career prospects of their mothers are especially limited.[19] In turn, those limitations also increase the rate of immature births, the perinatal mortality ratio, and the loss of infants at other times in the first year.

Women are by no means the sole contributors to problems that result in high perinatal mortality ratios. Fathers play a crucial role too, and if they smoke or drink heavily or are exposed to other toxins, they may incur genetic or other sperm damage that leads to defects in their offspring. Since fertilization is about as likely from defective as healthy sperm, the lifestyles, genetic make-up, and other attributes of fathers help determine the health of fetuses and infants, even though efforts to trace the source of fetal problems typically have focused on women.[20]

Low Birth Weight. The race differential in premature births also shows up in birth weight variations, with those of less than 2,500 grams considered to be "low birth weight."[21] In 1985–87, for example, 13 percent of black newborns but only 6 percent of the whites weighted less than 2,500 grams, and the percentage of black births at each birth weight under 2,500 grams was more than twice the percentage among whites. Conversely, only 26 percent of black newborns weighed at least 3,500 grams (7 lbs., 12 oz.), compared with 44 percent of white births. In addition, the median birth weight was 3,180 grams for blacks and 3,420 grams for whites. Low birth weight is doubly a problem for blacks because the proportion of underweight infants is greatest among teenagers, who produce a disproportionately large share of all black infants. The incidence of low birth weight is also significantly higher for blacks than whites at each specific age, from under 15 to 44–49. Furthermore, the percentage of low-birth-weight infants has decreased very little since 1978.[22]

Low birth weight is strongly implicated in perinatal mortality and infant mortality regardless of race, and it can even be considered the "best predictor of death in the first weeks of life."[23] In fact, one study showed that while only 6 percent of all newborns weighed less than 2,500 grams, they accounted for 70 percent of all infant deaths.[24] In turn, low birth weight plays an important role in several reported causes of death (e.g., congenital anomalies, pneumonia and influenza). Therefore, early and consistent prenatal care would either reduce the probability of low birth weight or help maximize the infant's survival potential despite being underweight. Unfortunately, many mothers and fetuses at risk either do not have such care available or receive it too late in the pregnancy.[25] Some women fail to use the prenatal care facilities that are available. The quality of prenatal care and access to it vary widely, and poor women may encounter insurmountable expenses and geographic and administrative barriers. Inaccessibility is costly, however, because only $400 worth of prenatal care could help ensure a healthy baby, whereas an unhealthy one can require $400,000 worth of extra services during its lifetime.[26] Many women and men also need to pursue healthier lifestyles so as not to produce children at risk of low birth weight, immaturity, and other problems. In fact, the leading causes of infant death are closely related to the lifestyles of both parents, especially women during pregnancy, and therefore fetal and infant well-being involves personal choices about smoking and alcohol and drug use.

The various risk factors in low birth weight are better known than the sociobiological mechanisms by which they actually operate to produce low birth weight. Therefore, some evidence suggests that genetic factors, such as a greater predisposition to certain illnesses among blacks than whites, may play a role in the racial differential in low birth weight. Nevertheless, when the socioeconomic risks that lead to early labor, retarded intrauterine development, and other proximate causes of low birth weight are held constant, most though not all of the racial difference disappears. Therefore, any genetic factors that are at work seem considerably less significant than whether or not one is disadvantaged socioeconomically.[27]

Unmarried Mothers. Infants conceived by unmarried mothers as a whole are more likely to die than are infants conceived by married women, even if the couple marries before delivery.[28] The disparity also results from differential prenatal care because an unmarried woman is much less likely than a married woman to receive such care. Marital status, however, is not a significant factor in perinatal or infant mortality among blacks and teenagers—groups with the highest percentages of unmarried births. Blacks and teenagers who are married are especially likely to have unemployed or poorly paid husbands, but being married keeps many from receiving public assistance.[29] Consequently, these poor married women often do not receive adequate or timely prenatal care, and they may suffer other disadvantages (e.g., malnutrition) that raise the infant mortality rate. Conversely, many unmarried black and teenage mothers have support units among kin and friends and often do receive public assistance, especially Aid to Families with Dependent Children (AFDC). Therefore, their incidence of prenatal care may parallel that among married teenagers and blacks, so their perinatal and infant mortality is no worse than among their married counterparts.[30]

Factors associated with differential perinatal mortality implicate not race per se, but poverty, medical neglect, drug abuse, and other problems to which black women are especially vulnerable because of their socioeconomic environment and which result in low birth weight and late or no prenatal care. Some are problems that better information can help individual black women and men overcome. Many, however, result from poor occupational opportunities, residence in inner-city ghettos, forces that hinder family formation and encourage family breakup, inadequate health-care services, and similar realities with roots in racism, sex discrimination, class polarization, and other structural flaws in the larger social system. Consequently, disproportionate numbers of the nation's black women are weakly integrated into the social system and are only marginal beneficiaries of it.[31] The higher perinatal mortality ratio reflects that marginality, especially the inability to buy the goods and services necessary for proper maternal, fetal, and infant health. It also reflects the attitudes and sense of responsibility parents have about childbearing and child care, and to the degree illegitimacy reflects irresponsibility, regardless of race, it is also likely to lead to poor prenatal care, high-risk behavior during pregnancy, inadequate child care after birth, and high infant losses. In part, these problems have a structural base, but they also involve a good deal of personal choice, much of it conditioned by peer influence and a type of social contagion.[32]

Two other differentials appear in the perinatal mortality ratios in Table 3-2— those by sex and metropolitan and nonmetropolitan residence. In the first case, perinatal mortality follows the general rule of sex variation: the mortality ratios of males are significantly higher than those of females, regardless of race. In the second case, the ratio is slightly higher for both races in nonmetropolitan than metropolitan places. Nonmetropolitan blacks have the worst situation, though it does reflect some countervailing forces in the relationships among residence, employment, poverty, family structure, and mortality.

Within Metropolitan Statistical Areas (MSAs), blacks are more likely to inhabit central cities, whereas whites are more apt to live in outlying suburbs. Central cities concentrate much of the nation's poor population of both races, while outlying parts of MSAs have relatively small poor populations. In 1989, for example, the poverty rate for whites in central cities of MSAs was 14 percent and that for blacks 33 percent. Only 26 percent of the nation's whites lived in central cities, however, compared with 57 percent of its blacks. On the other hand, whites residing in MSAs but outside central cities had a poverty rate of only 8 percent and blacks had one of 22 percent—both significantly lower than inside central cities. But 50 percent of the nation's white population lived in suburbs, compared with only 27 percent of its blacks. Thus, the heavy representation of blacks in central cities exposes a larger share of them to poverty conditions, poor health practices, and certain kinds of social disorganization, and the risks to their well-being are reflected in their relatively high perinatal mortality ratios.

Both whites and blacks in nonmetropolitan areas have higher poverty rates and perinatal mortality ratios than in the metropolitan sections, reflecting poor health care and similar problems in many small towns and other rural areas. Most doctors still avoid small towns and those who do locate there are increasingly likely to be specialists rather than general practitioners accessible to poorer women and children.

The black perinatal mortality ratio in rural areas would probably be even higher, however, if protective kinship and friendship patterns were not often better formed, more clear-cut, and tighter than those among rural whites. Consider that the proportion of black female-headed families is higher in nonmetropolitan places and so is the average number of children per family—factors that presumably should correlate with high perinatal losses. In nonmetropolitan places, however, black women who head families without a husband present are more likely to have at least one other adult present than are female-headed black families in metropolitan areas, including central cities and outlying districts, and they are more likely than white female-headed families in nonmetropolitan places to have the additional adult in the family. That person is often a young woman's mother and the family form that emerges, enmeshed in a larger social network, provides care and protection that depress the perinatal mortality ratio despite poverty and other negative forces.[33]

This family/friendship/community protection is one reason the nonmetropolitan perinatal mortality ratio among blacks fell 22 percent in the 1980s, compared with only 16 percent among metropolitan blacks. The change occurred despite a deteriorating economic situation for rural blacks (and other minorities) during the decade.[34] The change shows that the survival of fetuses and infants is affected by many factors in addition to the economic status of mothers; those other factors may counteract some of the impact that negative economic forces have on mortality. Nevertheless, in many rural areas financing for health care is inadequate and getting worse, and minority fetuses and infants are disproportionately victimized by that deficiency.

Neonatal and Postneonatal Mortality

Neonatal and postneonatal mortality rates separate infant deaths in the risky first 27 days after birth from deaths in the period of declining hazards during the rest of the first year. The relative importance of particular causes of death also changes as the probability of dying declines, although the factors that contribute to those causes may not change as much as previously thought. Furthermore, despite significant race and sex differentials in infant mortality, the percentages of deaths attributable to neonatal mortality are similar. In each race and sex category neonatal deaths account for about two-thirds of all infant deaths, postneonatal losses for the remaining one-third. Thus, in 1985–87 while the overall infant mortality rate among white males was 29 percent higher than among white females, 64 percent of the males and 66 percent of the females died during the neonatal period (see Table 3–4). The infant mortality rate among black males was 23 percent higher than among black females, but in both cases 66 percent of deaths occurred in the first 27 days. Even in the most extreme comparison, the infant mortality rate among black males was 155 percent higher than among white females, but 66 percent of the deaths in both groups were attributable to neonatal mortality. Consequently, the concentration of deaths in the first 27 days and the growing likelihood of survival throughout the rest of the first year are quite uniform.

On the other hand, risks are unevenly distributed by race and sex within the four-week neonatal period, when the death toll among infants in all categories is comparatively high. Table 3–5 shows neonatal mortality rates by sex and race and percentages of neonatals who died in specified time periods. The rates and percentages were computed so those deaths that occurred during periods of 7–13 days, 14–20 days, and 21–27 days are allocated to single days.

Among blacks and whites, males and females, the neonatal mortality rate at every age is significantly lower than at the age immediately preceding, and while the largest drop is between the first and second days of life, each succeeding day of survival is considerably safer than the previous day. As a result, a week-

Table 3-4
Infant, Neonatal, and Postneonatal Mortality Rates and Percent of All Infant
Deaths, by Race and Sex, 1985-87

Race and Sex	Infant Mortality		Neonatal Mortality		Postneonatal Mortality	
	Rate[a]	Percent	Rate[a]	Percent	Rate[a]	Percent
All races	1,036.0	100.0	671.2	64.8	364.8	35.2
White	896.1	100.0	580.0	64.7	316.1	35.3
Male	1,000.6	100.0	645.5	64.1	361.1	35.9
Female	779.0	100.0	511.0	65.6	268.0	34.4
Black	1,803.0	100.0	1,184.5	65.7	618.5	34.3
Male	1,982.4	100.0	1,302.0	65.7	680.4	34.3
Female	1,618.1	100.0	1,063.5	65.7	554.6	34.3

Source: National Center for Health Statistics, Vital Statistics of the
United States, 1985, 1986, 1987, Vol. II, Mortality, Part A (Washington,
DC: Government Printing Office, 1988-90), Sec. 2, Table 2.

[a]Deaths per 100,000 live births in specified group.

old infant has more than 38 times the survival prospects of one who is less than
a day old. Thus, many causes of infant mortality differ significantly between
the moment of birth and as little as a few days or a week after birth, not between
the entire neonatal and postneonatal periods.[35]

In one sense the record appears better for blacks than whites: a week-old black
infant has 51 times the survival potential of a newborn, whereas a week-old
white infant has only 34 times the prospects of a newborn. But even for those
who survive a week, the neonatal mortality rate is still much higher among
blacks, and part of their seeming ''advantage'' results from comparing very high
losses on the first day with subsequent losses: 63 percent of all black neonatal
deaths occur during the first day, compared with 56 percent of the white deaths.
This is when factors such as age of the mother, her prior history of fetal or infant
deaths, her lifestyle, the efficacy of prenatal care, birth weight, and the father's
genetic contribution strongly affect survival. In turn, they are affected by the
mother's education, her marital status, and certain aspects of her economic
level.[36] For example, inner-city blacks in their childbearing years are less likely
to be employed than older workers, and the higher rates of unemployment
contribute to higher rates of nonmarriage, divorce and desertion, one-parent
family formation, and other forces that compromise health maintenance, prenatal
care, and related conditions that influence infant mortality. These conditions then
interact to condense a higher proportion of black than white deaths into the first
24 hours, though even the negative conditions are partly counterbalanced by

Table 3-5
Neonatal Mortality Rate Per Day of Age, and Percent of All Neonatal Deaths, by Race and Sex, 1985–87

Race and Sex	Age										
	0–27 Days	0–1 Day	1 Day	2 Days	3 Days	4 Days	5 Days	6 Days	7–13 Days[a]	14–20 Days[a]	21–27 Days[a]
Rate[b,c]											
All races	671.2	387.5	68.2	41.3	24.6	15.9	12.2	10.1	7.4	4.9	3.6
White	580.0	323.7	59.9	37.1	22.8	15.0	11.2	9.4	6.9	4.4	3.2
Male	645.5	352.8	70.2	43.2	26.1	16.8	12.7	11.1	7.7	5.0	3.4
Female	511.0	293.0	49.0	30.6	19.3	13.2	9.6	7.7	6.0	3.8	2.9
Black	1,184.5	742.5	114.5	65.7	36.3	21.1	18.3	14.6	10.7	7.5	6.2
Male	1,302.2	819.0	128.1	75.4	41.5	23.4	19.8	15.7	11.4	7.8	6.4
Female	1,063.5	663.6	100.5	55.8	31.0	18.7	16.9	13.5	10.0	7.3	6.1
Percent[c]											
All races	100.0	57.7	10.2	6.1	3.7	2.4	1.8	1.5	1.1	0.7	0.5
White	100.0	55.8	10.3	6.4	3.9	2.6	1.9	1.6	1.2	0.8	0.5
Male	100.0	54.7	10.9	6.7	4.0	2.6	2.0	1.7	1.2	0.8	0.5
Female	100.0	57.3	9.6	6.0	3.8	2.6	1.9	1.5	1.2	0.7	0.6
Black	100.0	62.7	9.7	5.6	3.1	1.8	1.5	1.2	0.9	0.6	0.5
Male	100.0	62.9	9.8	5.8	3.2	1.8	1.5	1.2	0.9	0.6	0.5
Female	100.0	62.4	9.4	5.2	2.9	1.8	1.6	1.3	0.9	0.7	0.6

Source: National Center for Health Statistics, Vital Statistics of the United States, 1985, 1986, 1987, Vol. II, Mortality, Part A (Washington, DC: Government Printing Office, 1988–90), Sec. 2, Tables 2, 3.

[a] Figures are allocated rates per day.

[b] Death per 100,000 live births in specified group.

[c] Horizontal rows do not cumulate to 100 percent because of per-day allocation in last three columns.

public assistance programs and other local, state, and federal efforts to reduce neonatal deaths among the poor.

Deaths of black neonatals are somewhat less concentrated than those of whites in days 2–6, partly because defective white infants are more apt to be kept alive a few days longer than black ones; whites as a whole have superior access to the technology and personnel that help sustain even the most precarious lives. Therefore, some seriously defective white infants would have been stillbirths without those protections but manage to survive birth and the first, second, or more days. For blacks as a group, the winnowing process is harsher and earlier. By the end of the first week, however, the percentage of infant deaths attributable to each successive day is about the same for blacks and whites, males and females. Consequently, while the neonatal mortality rate is 130 percent higher for blacks than whites on the first day, by the end of a week it is only 55 percent higher. The heavier initial losses indeed may be harsh, but the survivors have a better chance for continued survival, and the racial gap narrows. This phenomenon has some features of the racial crossover among the elderly, though among infants the convergence neither closes the gap nor becomes a crossover. In any case, a large share of the white advantage in infant mortality is due to the race differential in very early neonatal mortality.

DIFFERENTIALS BY SEX

Larger percentages of males than females die at every age, including infancy. Are there ages within the first year, however, when the survival potential of boys relative to girls is better than at other ages? Does the sex mortality differential precede birth and show up in fetal deaths? Do male and female infants exhibit similar or different patterns of loss among the major races? The sex mortality ratio, derived for various stages during the first year of life, provides some answers.

Males are significantly more likely than females to die during the entire period from late gestation to the first birthday. Even among fetal deaths (stillbirths) about 107 males are lost for each 100 females, and since that is somewhat higher than the sex ratio of live births (105), even gestation is riskier for males than females. But the situation differs markedly by race. Among whites the risks result in a loss of about 105 male fetuses for each 100 females—the same as the sex ratio of live births. Among blacks, however, about 112 male fetuses die per 100 female fetuses, which is substantially higher than the sex ratio at birth (103). Therefore, factors that threaten black fetuses more than white ones take an exceptionally high toll of males.

The sex mortality ratio rises during the perinatal period, reflecting dispropor- tionate losses of males after 28 weeks gestation and into the first six days of life (see Table 3–6). At that stage too the sex ratio is higher for blacks than whites. It is even higher in the neonatal period, so the first several days of life are especially dangerous for boys. Moreover, the sex mortality ratio remains rela-

Table 3–6

Sex Mortality Ratio for Perinatal, Neonatal, and Postneonatal Mortality, by Age and Race, 1985–87

Type of Mortality and Age	All Races	White	Black
Perinatal[a]	113.4	112.8	116.4
Neonatal			
0–27 days	124.5	126.3	122.4
0–1 day	120.7	120.4	123.4
1 day	137.6	143.3	127.5
2 days	139.7	141.2	135.2
3 days	134.4	135.2	133.9
4 days	128.1	127.3	125.1
5 days	127.1	132.2	117.2
6 days	136.5	144.2	116.3
7–13 days	125.2	129.5	113.9
14–20 days	123.3	132.1	107.3
21–27 days	116.7	117.2	105.2
Postneonatal			
28–59 days	131.5	137.6	120.7
2 months	137.2	140.4	129.9
3 months	142.6	148.6	135.6
4 months	132.6	134.6	126.9
5 months	129.4	134.0	121.9
6 months	122.4	122.8	115.4
7 months	124.8	132.2	106.8
8 months	117.1	120.6	107.5
9 months	113.0	110.4	121.8
10 months	103.0	109.4	86.6
11 months	105.3	102.3	121.3
Under 1 year	126.6	129.2	122.5

Source: National Center for Health Statistics, Vital Statistics of the United States, 1985, 1986, 1987, Vol. II, Mortality, Part A (Washington, DC: Government Printing Office, 1988–90), Sec. 2, Table 2.

[a]NCHS Definition I. (See Table 3-2.)

tively high until after the first six months or so. Finally, by the end of the first year, while death continues to discriminate against boys, it does so at a much lower rate than in the first days, weeks, and even months. At times during those early periods the death rate among boys is as much as 40 percent higher than among girls, and not only are infant deaths grouped within a short period after birth, so is the disproportionate loss of males.

Despite heavy losses of black males as fetuses and newborns, the sex discrepancy becomes smaller for blacks and whites into the neonatal and postneonatal stages. Even allowing for data deficiencies, at most ages after the first day blacks experience more nearly equal losses of boys and girls than do whites. At some ages the racial differentials are large, at others they are small, but at nearly every age the sex mortality ratio is closer to parity among blacks than whites.

Maternal Mortality

Quite a different sex-specific death rate is worth noting in this section because it stems from the birth process and is closely related to fetal and infant maladies. Like the infant mortality rate, it is a fairly sensitive indicator of the well-being of a population. The index is the maternal mortality rate, or the number of deaths of women from complications of pregnancy, childbirth, and the puerperium per 100,000 live births.

Historically, maternal mortality has fallen dramatically, from 728 deaths per 100,000 live births in 1915–19 to only 7.4 in 1986–88. This is far more rapid than the decline in general mortality and infant mortality and it reflects a range of lifesaving successes. Nevertheless, the black/white differential persists, and in 1986–88 maternal mortality rates were 5.3 for whites and 17.5 for blacks. The rate for blacks is lowest under age 20, after which it rises steadily, and that for whites is lowest in the ages 20–24, after which it also rises consistently. Thus, the tendency for black women to bear children at younger ages than white women helps keep the number of black maternal deaths lower than it might be otherwise, whereas the more pronounced delayed childbearing pattern among white women puts more of their births into the ages with greater risk of maternal mortality. For example, about 56 percent of all black births but only 38 percent of the white ones take place under age 25. About 5 percent of the black births but 8 percent of the white ones occur to women 35 and over. Moreover, the risk of an unfavorable pregnancy outcome (e.g., stillbirth, infant death) tends to increase with age beyond 35, though not dramatically.[37] The risk of Down's syndrome, but not other birth defects, also increases significantly with the mother's age.

On balance, the long-term decline in maternal mortality is a success story for both races. Blacks, however, still have a rate more than triple that among whites, and births to older women are more likely to result in maternal mortality, stillbirth, infant mortality, and certain birth defects, but only to a minor degree. Delayed childbearing is likely to persist in the foreseeable future, but it should continue to have only a limited impact on the rate of maternal mortality and infant deaths.

DIFFERENTIALS BY RACE AND ETHNICITY

Earlier parts of this chapter focused on differentials between blacks and whites in perinatal, neonatal, postneonatal, and even fetal mortality. It remains to examine other aspects of that racial relationship, several other racial groups, and the Hispanic population, although data for all groups except blacks and whites are fragmentary and subject to substantial errors that lead to underestimates of infant mortality. The data for the Spanish-origin population and its several components are especially lacking in detail and are available for only half the states and the District of Columbia. Those units contain a large majority of the persons with Mexican and Puerto Rican backgrounds, but only somewhat more than a third of Cuban-origin persons, largely because Florida is not included in the list.[38] In addition, the data seriously underestimate infant deaths among Filipinos, Native Americans, and even Japanese and other groups. Despite the methodological problems, some conclusions are possible (see Table 3–7).

Compared with all other racial and ethnic groups in the United States, the infant mortality rate among blacks is extremely high. No other group has a reported rate that comes close, and even Native Americans, many of whom also suffer severe socioeconomic disadvantages, have a reported infant mortality rate considerably lower than that among blacks. Other groups also appear to fare well, compared not only with American blacks, but also with whites. Thus, the Chinese, Japanese, and Filipino populations have comparatively low reported infant mortality rates, and even other Asians and Pacific Islanders are below the national average. Hispanics also fare relatively well: even allowing for data limitations, that group as a whole has an infant mortality rate close to the national average and much lower than the rate among blacks.[39]

Within the Hispanic population the infant mortality rate is lowest among Cubans, somewhat higher among Mexicans, and somewhat higher still among Puerto Ricans, who are the poorest Spanish-origin population. As noted in Chapter 2, the death rate is relatively high for Puerto Ricans because of the large group living in New York City; Puerto Ricans elsewhere generally have much lower death rates, including infant mortality.

Income and poverty play a role in the racial and ethnic differences because parents who can neither afford adequate health care privately nor pay premiums for health insurance often lack access to the personnel and facilities that save fetuses, infants, and mothers. Therefore, the fact that blacks have a higher poverty rate than all other groups listed in Table 3–7 negatively affects infant mortality.[40] In some comparisons poverty differentials are pronounced: black families have 2.5 times the poverty rate of Chinese families, 6.3 times the rate of Japanese families, and 4.3 times the rate of Filipino families. In other comparisons the economic discrepancies are not sufficient to explain much of the mortality differential. For example, blacks have a poverty rate only 17 percent higher than that among Hispanics, but the black infant mortality rate in 1985–87 was 120 percent greater than that in Hispanic families. In addition, while the median family income of blacks is much lower than those of Chinese, Japanese, and Filipinos, it is not much below those of various Pacific Islanders or Hispanics, and yet large mortality differentials persist. Although the poverty rate of Native Americans is only slightly lower than that of blacks and their median income is

Table 3–7
Infant, Neonatal, and Postneonatal Mortality Rates, by Race and Ethnicity, 1985–87

Race and Ethnicity	Infant Mortality[a]	Neonatal Mortality[a]	Postneonatal Mortality[a]
All races	10.4	6.7	3.7
White	8.9	5.8	3.1
Black	18.0	11.8	6.2
American Indian	9.8	4.8	5.0
Chinese	5.0	2.7	2.3
Japanese	4.3	2.8	1.5
Filipino	4.7	3.2	1.6
Other Asian or Pacific Islander	6.2	4.9	2.3
Other races	4.5	3.0	1.5
Hispanic	8.2	5.2	2.9
Mexican	7.9	5.0	2.9
Puerto Rican	8.0	4.8	3.2
Cuban	6.9	4.8	2.1
Other Hispanic	9.1	6.1	3.0
Non-Hispanic	10.0	6.3	3.7
White	8.6	5.5	3.1
Black	16.7	10.5	6.3

Source: National Center for Health Statistics, Vital Statistics of the United States, 1985, 1986, 1987, Vol. II, Mortality, Part A (Washington, DC: Government Printing Office, 1988–90), Sec. 2, Tables 4, 19.

[a]Deaths per 1,000 live births in specified group.

well below that of blacks, the infant mortality rate is 84 percent higher for blacks than for Native Americans.[41]

Black newborns and infants receive much less adequate health care than whites, partly because of greater parental poverty and educational deficiencies. In 1988, for example, black infants were two to three times less likely than whites to have had any kind of medical checkup or any immunization, and the attention they did get was apt to be in a clinic or hospital emergency room, not from a private practitioner. Infants from the inner city, regardless of race, were about twice as likely as suburban children to receive their routine care in a clinic or emergency room, usually as crisis treatment rather than prevention. About 47 percent of black infants but only 16 percent of whites received their care at a clinic, and while many clinics perform valiantly, they are notoriously under-funded and their patient loads are exceptionally heavy.[42] One reason black infants are more likely to visit clinics than private physicians is their parents' lesser ability to pay for health insurance: only about 56 percent of the nation's blacks but 80 percent of the whites have private health insurance coverage. Moreover, less than 40 percent of families with incomes below $10,000 have coverage, compared with 95 percent of those with incomes of $50,000 or more.[43] Medicaid does narrow the coverage gap somewhat, but smaller percentages of blacks than whites and fewer poor than affluent children still have any kind of health coverage, and the differences are partly responsible for high infant mortality rates among blacks. Even so, the medical care and insurance experience of Hispanics is not very different from that of blacks, and yet the Hispanic infant mortality rate is considerably lower.[44]

In all these comparisons the black population stands alone with its inordinately high infant mortality rate, so while income and poverty play a part, other strategic factors also are at work.

Education is one factor, but even its role varies. The percentages of high school and college graduates are significantly higher among Chinese, Japanese, Filipinos, and whites than among blacks. Education does help people understand proper diet, appropriate health care, timely and consistent medical attention during pregnancy, and the dangers of certain lifestyle factors, such as smoking and drug and alcohol abuse. In addition, higher levels of education are correlated with later age at childbearing, a lower incidence of births to unmarried women, and other aspects of reproduction that reduce the rate of premature and underweight births and thereby reduce fetal and infant mortality. But the proportion of Native Americans who graduated from high school is barely higher than that among blacks and the percentage of college graduates is lower, so the educational difference between those races is not enough to account for the large difference in infant mortality. Moreover, the percentages of high school and college graduates are lower among Hispanics than among blacks, but the large infant mortality gap persists between them.

The unemployment rate among blacks is more than double that among whites, Japanese, and Chinese; it is nearly twice that among Filipinos and one-third higher than among Hispanics. Unemployment can lead to poverty, isolation in inner cities and other places with dangerous lifestyles, inadequate diet and health care, exposure to communicable diseases, and resentment and abuse of unwanted infants whose presence exacerbates deprivation and limits a parent's employment prospects. These and other consequences of unemployment may result directly or indirectly in higher rates of fetal and infant mortality. Yet the unemployment rate among Native Americans is considerably higher than among blacks, so unemployment by itself does not explain much of the mortality differential. What does?

One condition is much more prevalent among blacks than in any other race and ethnic group, and while it is partly a product of factors already discussed and interacts with them in complex ways, it may be the single most important factor implicated in the high black infant mortality rate. That is, the black family is less likely than ever to consist of a married couple and more likely to be headed by a woman with no husband present—far more so than in any other race and ethnic group. In 1988, for example, married-couple families made up only 51 percent of all black families, compared with 70 percent among Hispanics and Native Americans, 83 percent among whites, and even higher proportions among Chinese, Japanese, and Filipinos. Conversely, female-headed families were 43 percent of all black families, but only 23 percent among Hispanics and Native Americans, 13 percent among whites, and even lower percentages among the other races.

Single parenthood by itself does not raise infant mortality rates, and many mother-only families are safe places for fetuses to develop and infants to be born and to flourish. Compared with married-couple units, however, the odds are not favorable because the socioeconomic problems that often accompany single parenthood aggravate other conditions that endanger new lives. Some of the interactions are as follows.

1. The likelihood of unemployment, low income, and poverty is greater among female-headed families and so, therefore, is the chance that a mother cannot afford a proper diet, prenatal care, or necessities for a newborn.

2. The one-parent family is more a phenomenon of central cities than suburbs or rural areas, and inner-city hazards endanger the young. They include the complex, seemingly intractable problem of drug dealing and use, high rates of violence, overburdened clinics and other health-care facilities and personnel, and poor employment prospects. In fact, several demographic forces have operated in the past few decades to concentrate more and more of the poor black population into enclaves where mortality rates at any age, but especially infancy, are particularly high. These forces include a later and lesser drop in black than white fertility, a relatively high rate of childbearing among very young women, and metropolitanization of poor blacks and suburbanization of upwardly mobile blacks. In turn, concentrations of poor blacks in central cities contribute further to homogeneous young cohorts in their childbearing years, teen pregnancy, subculture shaped by poverty, employment problems, substance abuse, rising percentages of mother-only families, low-birth-weight newborns, and, ultimately, a high rate of infant mortality.[45]

3. Overall family structure and stability are weaker where the mother-only family is most common because rates of divorce, desertion, and births out of wedlock all are higher. The structural weakness may leave the mother and infant without an appropriate support network of kin and friends to provide financial assistance, advice about prenatal care, day care, and other amenities that enhance infant survival. As noted earlier, the relative strength of this support network differs between metropolitan and nonmetropolitan populations. In addition, many young fathers default on financial and child-care responsibilities, leaving young mothers to generate income, care for children physically, and socialize them. Inadequate physical care, child abuse, and even emotional deprivation contribute to infant mortality. These problems are affected by whether or not the child was wanted: 22 percent of children born to black mothers were unwanted at conception, compared with 8 percent among whites, although the proportions have fallen for both races since the early 1970s.[46]

4. When a mother-only family forms, especially because of birth out of wedlock, the young woman usually quits formal education, and the resultant deficiencies in skills and sophistication impede employment and/or successful prenatal and infant care and parenting, thus raising the risk of death.

The two primary victims of these conditions are mother and infant, and Margaret Boone reports a profile of the black high-risk, inner-city pregnant woman that underscores this victimization. The mother, who typically delivers an underweight infant, is especially likely to abuse nicotine and alcohol, as is the child's father; she is apt to receive little or no prenatal care and to have a history of poor pregnancy outcome, reflected in repeated abortions, loss of infants, and frequent stillbirths. She is also likely to have a history of one or more chronic health problems, especially hypertension, and to be in poor general health. In addition, women from the inner city who do have underweight infants usually become pregnant quite young; they are underweight themselves; their pregnancies are closely spaced; they are ineffective users of contraception. Such women often lack regularized and dependable social supports and many are psychologically depressed because of unsatisfactory relationships; they are prime victims of physical and psychological abuse. Women afflicted by these conditions are apt to contribute repeatedly to the roster of fetal, perinatal, and infant deaths and, therefore, to produce far more than their proportionate share of the infant mortality problem in the black population.[47]

These are only parts of the complex constellation of biological, social, economic, and other factors that cause high infant mortality rates. The mother-only family is much more common among blacks than other racial and ethnic groups in the United States. But it is a symptom of these other factors, and rather than seek a single cause or correlation, we would do better to recognize the many conditions that interact to keep black fetal and infant mortality rates high. The long-term solution, therefore, is concomitantly complex and involves fundamental social restructuring and equitable realignment of opportunities for blacks and whites, especially young women, within the social system, and whatever steps are necessary to enhance individual responsibility. A more short-term solution is to institute a national health program and to make its protections available to all Americans.[48]

The problems just discussed thwarted a 1979 national goal for blacks. In that year overall infant mortality was 13.1 and the Surgeon General's goal was to lower it 35 percent to 9.0 by 1990.[49] The goal has been met for whites but not for blacks or, therefore, for the nation as a whole. Black infant mortality in 1990 was still nearly twice the target rate, and the pace of decline was considerably slower in the 1980s than in the 1970s. High black infant mortality remains one of the nation's most persistent problems.

GEOGRAPHIC DISTRIBUTION OF INFANT MORTALITY

Infant mortality rates are much higher in some regions, divisions, states, and counties than others, often because of racial distribution and other differences in population composition, or because of substantial variations in the adequacy of maternal and infant health services.[50]

Geography and Race

Infant mortality variations among the four large regions are significant. The South fares the worst and the West the best, with a difference of 19 percent (see Table 3–8). The differences among the divisions are somewhat greater, with the highest rates in the South Atlantic and East South Central divisions, and the lowest in New England, for a difference of 37 percent. Even that variation pales, however, beside the black/white differential in every region and division and in each state with enough black infant deaths to enable comparisons. Given the importance of race in determining differences in infant mortality, the focus on geographic variations again points to a concern with black/white variations.

In the Northeast the black/white infant mortality ratio is 209, since that region has a lower infant mortality rate for whites but a higher one for blacks than any other region, including the South. The racial gaps are especially large in Massachusetts, Connecticut, New Jersey, and Pennsylvania—states with some of the nation's most urbanized areas and worst problems for inner-city blacks. Even in the West, where many blacks fare better than elsewhere in the nation, the black/white mortality ratio is 174; in the Midwest it is 190, in the South 197. By comparison, the regional differences among blacks in infant mortality are relatively small, just as they are among whites. Thus, infant mortality rates are far more similar *within* each racial group than they are *between* them, regardless of the geographic unit, for despite the progress black women have made since 1960 in education, job status, and other areas, by many criteria they lost ground in the 1980s.

Patterns in the nine census divisions are similar to those in the four regions though the gaps are larger, with black/white infant mortality ratios ranging from 162 in the Mountain division to 226 in New England.

Differences by States

Patterns of infant mortality in the divisions reflect those in the component states, so most conclusions about the larger units apply to the smaller ones. As expected, racial composition plays an important role. There are notable state variations in infant mortality rates among whites, from 7.5 in Massachusetts to 11.6 in the District of Columbia, just as there are among blacks, from 13.0 in Rhode Island to 23.5 in the District of Columbia. But the race differential within each state is much greater than the state-to-state variations within each race. In 1985–87, 41 states and the District of Columbia had sufficient black infant deaths to compute a rate. In 16 of those 42 units the infant mortality rate of blacks was at least twice that of whites, and in every state except Washington the black rate was at least half again as high as the white rate. The racial mortality ratio was

Table 3–8

Infant Mortality Rate, by Race, Region, Division, and State, 1985–87

Region, Division, and State	Rank[a]	All Races[b]	White[b]	Black[b]	Black/White Ratio
United States	...	10.4	9.0	17.6	195.6
Northeast	...	10.0	8.6	18.0	209.3
New England	...	8.6	8.0	18.1	226.3
Maine	5	8.7	8.8	c	c
New Hampshire	4	8.7	8.7	c	c
Vermont	7	9.0	8.9	c	c
Massachusetts	1	8.3	7.5	17.6	234.7
Rhode Island	3	8.7	8.4	13.0	154.8
Connecticut	15	9.2	7.9	19.6	248.1
Middle Atlantic	...	10.5	8.8	18.0	204.5
New York	33	10.7	9.2	16.7	175.8
New Jersey	22	9.9	8.0	18.4	230.0
Pennsylvania	30	10.5	8.8	20.8	236.4
Midwest	...	10.4	9.0	17.1	190.0
East North Central	...	10.8	9.1	20.1	220.9
Ohio	25	10.1	9.0	16.5	183.3
Indiana	35	10.8	9.7	20.1	207.2
Illinois	44	11.8	9.3	21.4	230.1
Michigan	40	11.2	8.9	22.2	249.4
Wisconsin	8	9.0	8.3	17.1	206.0
West North Central	...	9.5	8.8	17.1	194.3
Minnesota	6	8.9	8.6	16.5	191.9
Iowa	10	9.0	8.9	14.9	167.4
Missouri	29	10.4	9.1	17.7	194.5
North Dakota	2	8.6	8.3	c	c
South Dakota	39	11.0	9.1	c	c
Nebraska	18	9.4	8.8	18.0	204.5
Kansas	14	9.2	8.7	16.3	187.4
South	...	11.2	9.1	17.9	196.7
South Atlantic	...	11.8	9.2	18.6	202.2
Delaware	47	12.7	10.2	21.0	205.9
Maryland	42	11.7	9.1	18.1	198.9
District of Columbia	51	20.4	11.6	23.5	202.6
Virginia	37	10.9	8.8	18.0	204.5
West Virginia	26	10.3	9.8	21.3	217.3
North Carolina	43	11.7	9.4	17.8	189.4
South Carolina	50	13.4	9.8	19.2	195.9
Georgia	46	12.6	9.8	18.3	186.7
Florida	38	10.9	8.7	18.3	210.3

Table 3-8 (continued)

Region, Division, and State	Rank[a]	All Races[b]	White[b]	Black[b]	Black/White Ratio
East South Central	...	11.8	9.4	18.3	194.7
Kentucky	27	10.3	9.8	15.3	156.1
Tennessee	41	11.4	8.9	19.5	219.1
Alabama	48	12.7	9.7	18.7	192.8
Mississippi	49	13.3	9.4	17.8	189.4
West South Central	...	10.1	8.8	16.2	184.1
Arkansas	34	10.8	9.5	15.1	158.9
Louisiana	45	11.9	8.5	17.3	203.5
Oklahoma	28	10.3	10.2	16.6	162.7
Texas	19	9.5	8.6	15.5	180.2
West	...	9.4	9.1	15.9	174.7
Mountain	...	9.5	9.3	15.1	162.4
Montana	23	10.0	9.3	c	c
Idaho	31	10.7	10.7	c	c
Wyoming	36	10.9	10.8	c	c
Colorado	16	9.3	9.0	16.5	183.3
New Mexico	17	9.4	9.2	17.3	188.0
Arizona	20	9.5	9.1	14.3	157.1
Utah	9	9.0	9.1	c	c
Nevada	12	9.1	8.8	15.3	173.9
Pacific	...	9.3	9.0	16.0	177.7
Washington	24	10.1	10.0	13.8	138.0
Oregon	21	9.9	9.9	18.0	181.8
California	13	9.2	8.8	16.2	184.1
Alaska	32	10.7	9.4	13.3	141.5
Hawaii	11	9.0	8.1	15.5	191.4

Source: National Center for Health Statistics, Vital Statistics of the United States, 1985, 1986, 1987, Vol. I, Natality (Washington, DC: Government Printing Office, 1988-89), Sec. 1, Tables 51, 53; Vol. II, Mortality, Part B (Washington, DC: Government Printing Office, 1988-89), Sec. 8, Table 1.

[a]Rates for tied states computed to sufficient decimal places to rank.

[b]Deaths per 1,000 live births in specified group.

[c]Insufficient data.

highest in Michigan (249), lowest in Washington (138). Large racial disparities prevail even in Massachusetts, New Jersey, Wisconsin, Minnesota, and other states where the overall infant mortality rate is well below the national average. There are also striking geographic variations within each race, especially at the local level, and differential infant mortality is partly a state and community problem because of wide differences in health care. Until 1981, under federal mandate and with federal funds, state and local health departments provided certain kinds of free or low-cost health services to poor women. The 1981

Omnibus Budget Reconciliation Act replaced those specified programs with the Maternal-Child Health Block Grant, which dropped most federal standards and reporting requirements. The states then decided the kinds of programs that they would provide and each defined who was eligible. Now, each state has its own way to meet the health needs of poor women and children, and the earlier uniformity has given way to wide diversity in the quality of and expenditures for particular health measures. Moreover, the money appropriated for the block grant in 1981 did not increase much thereafter, and even the initial amount was less than that appropriated to the states under the old system. Some states supplement the block grant generously and have very good maternal and child health care. Others are much more parsimonious and their services are poorer than in 1981.[51] Consequently, infant mortality rates vary significantly from state to state and locale to locale, though virtually every governmental unit is struggling with increasing financial constraints, and health programs for the poor are tempting targets for cuts.[52] The Women, Infants, and Children (WIC) program and other federal efforts were much more effective than the fragmented approach of the 1980s and early 1990s.

The country's relatively high infant mortality rate is largely a national problem, despite the tendency to leave it to the states and local areas, though local variations in health do account for many geographic differences in infant mortality. The high rates among blacks almost everywhere show that black women have suffered more than their fair share of disadvantages because of changes in the health-care system. Obviously, this does not apply to the entire black population: the use of aggregate data does not always allow adequate distinctions between the black upper and middle classes on the one hand and the poor on the other. The data do reflect, however, the higher incidence of poverty, maternal neglect, violence, and other problems in the black population and the dangers those problems create for poor infants no matter what their race. Mortality analysis should not perpetuate racial stereotypes, but neither should it ignore the fact that blacks are the most victimized by a high infant mortality rate and its many causes, regardless of region, division, state, community, or even neighborhood. Therefore, the high rate among blacks reflects structural conditions that affect the whole society and, to varying degrees, each of its parts, including individuals.

County Comparisons: High-Risk and Low-Risk Areas

The state is still a relatively large unit to study the geographic distribution of infant mortality, which differs widely among much smaller units. Since state and national programs and other conditions affect local areas, which often bear the primary responsibility for whatever health care is available, the county is the most practical subdivision to pinpoint areas of high and low infant mortality. In this case, the data are for the six-year period 1983-88, partly to minimize the effect of annual fluctuations, but largely to accumulate a sufficient number of infant deaths and births to compute infant mortality rates that reflect local conditions fairly accurately. Even so, among the nation's more than 3,100 counties over 400 had so few infant deaths that no data can be reported. Most are sparsely inhabited rural counties in such states as Nebraska, Montana, the Dakotas, and Texas. Moreover, the number of blacks and their infant deaths are so small in so many counties that Figure 3–1 shows the distribution of infant mortality for all races collectively, although other studies do distinguish between the races in specific local areas.[53] Finally, data quality varies from county to county, although the rates are probably accurate enough to portray the principal geographic patterns. The map and some data not reflected on it enable several observations.

Infant mortality rates are still far above the national average in large parts of the South, despite progress in the region as a whole. The counties with large percentages of blacks, especially poor women heading families without a husband present, have abnormally high rates. One area with very high rates stretches south and west from eastern North Carolina through South Carolina, Georgia, Alabama, Mississippi, and on to northern Louisiana and into Arkansas and parts of Missouri. It includes sizable parts of Virginia, West Virginia, Tennessee, and Kentucky. Some of these counties are in old plantation sections, where historically a small landowning elite and a large mass of landless laborers made up a two-class system, though a small middle class was sometimes present. Other high-mortality counties are in the clay hills of north Georgia, the piney woods sections of several states, and other poor areas whose people are inadequately served by health-care professionals. Many of these areas still have too few industries to provide an adequate tax base for local services; many jobs pay poorly and many people cannot afford health insurance or other protections. Organized labor also is generally weak in these places, and much of the income from the agriculture that remains is low. This large southern area is quite diverse and parts of it do not have sizable black populations, but much of it does have a relatively high percentage of female-headed families, many in poverty.

Infant mortality rates are also high in parts of the upper Midwest and in Oklahoma and northern and western Texas. But highly urbanized states too, such as Illinois, Michigan, New York, and California, have numerous counties with above-average infant mortality rates. Many of these counties lie in the heart of MSAs, and their fairly high rates reflect the problems of many central cities (e.g., Detroit). Some of those counties are bordered by others in the same MSA with extremely low rates; they are generally the middle- and upper-class suburban counties with comparatively high levels of living, adequate health facilities, and well-educated populations that understand and can afford good care for mothers and infants. Some examples of this pattern are the counties containing and surrounding Boston; Newark, Jersey City, and Trenton; Pittsburgh, Philadelphia, and some other Pennsylvania cities; Cleveland, Akron, Columbus, and Cincinnati; Milwaukee; Gary and Indianapolis; and Chicago. The District of Columbia

Figure 3–1
Infant Mortality Rate, by County, 1983–88

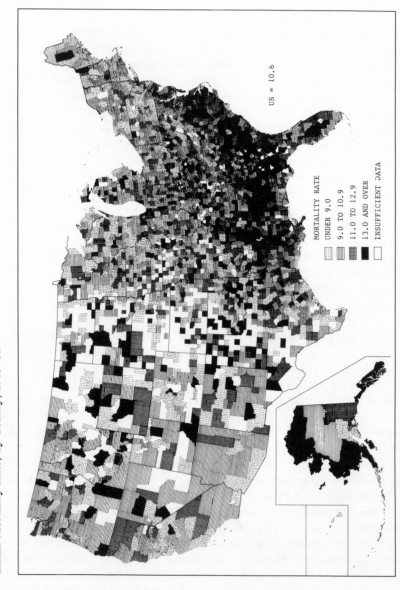

MORTALITY RATE

UNDER 9.0
9.0 TO 10.9
11.0 TO 12.9
13.0 AND OVER
INSUFFICIENT DATA

US = 10.6

Source: National Center for Health Statistics, *Vital Statistics of the United States,* 1983, 1984, 1985, 1986, 1987, 1988, Vol. I, *Natality* (Washington, DC: Government Printing Office, 1987–90), Sec. 2, Table 1; Vol. II, *Mortality,* Part B (Washington,

is a prime example. In other states, particularly in the South, some counties containing central cities of MSAs have comparatively low infant mortality rates, while less urbanized bordering counties have relatively high ones.

Fairly low infant mortality rates appear in most of New England, with some exceptions; in much of New York State outside New York City; and in numerous counties in the Midwest, many near major cities. The counties in several northwestern states fare well, as do some in the Southwest. Otherwise, every state even those with a dismal overall record of infant mortality, has its scattered islands of success in saving infant lives. They are generally counties with good prenatal care programs that reach a large share of pregnant women, counties with gynecologists and obstetricians readily available, and those with relatively few cases of sexually transmitted diseases and drug and alcohol abuse among mothers or fathers. They also have comparatively few underweight infants and other conditions that contribute to high infant mortality.

On balance, certain discernible patterns correlate with either high or low infant mortality rates in the counties. The urban core counties have relatively high rates, suburban counties have very low ones, and extremely rural counties tend to have the highest rates of all, though more consistently in the South than elsewhere. Despite the patterns, infant mortality rates in the counties are highly individualized according to local economies, employment conditions, median incomes, health-care facilities, and other factors that vary widely over short distances. Where the black population is sizable, a county's record of infant mortality is influenced by differential health care available to blacks and whites, and in all too many counties whites fare well and their infant mortality rates are quite low, while the reverse is true for the black population.

Urban and Nonurban Differentials

Variation in infant mortality rates by urban and nonurban residence is also a form of geographic distribution. Some differences in perinatal mortality between metropolitan and nonmetropolitan populations, discussed earlier, are consistent with differences in infant mortality by residence.

The rates in Table 3–9 are for the urban population, separately in metropolitan and nonmetropolitan counties, and for the "nonurban" population. The urban group resides in places of 10,000 or more, the nonurban group outside those places. The census definition of urban, however, includes all places of 2,500 or more, so the population in places of 2,500–10,000 is not strictly nonurban and it certainly is not rural-farm. Thus, the nonurban category includes persons in small cities, towns, villages, and the open country. Nevertheless, that category enables useful comparison with the nation's great metropolitan centers. The patterns also vary by race.

Whites have the highest infant mortality rates in the urban parts of nonmetropolitan counties, while blacks have the highest rates in the urban parts of metropolitan counties. Thus, the inner cities of large metropolises are the riskiest

Table 3–9
Infant, Neonatal, and Postneonatal Mortality Rates, by Residence and Race, 1985–87

Type of Mortality and Residence	All Races	White	Black
Infant[a]			
All urban[b]	11.1	9.3	18.2
Metropolitan urban	11.1	9.2	18.3
Nonmetropolitan urban	10.6	9.6	16.4
All nonurban	9.4	8.6	17.5
Metropolitan nonurban	8.9	8.3	17.5
Nonmetropolitan nonurban	10.1	9.1	17.5
Neonatal[a]			
All urban[b]	7.2	6.0	12.0
Metropolitan urban	7.3	6.0	12.1
Nonmetropolitan urban	6.5	5.9	10.3
All nonurban	6.0	5.5	11.4
Metropolitan nonurban	5.8	5.4	12.0
Nonmetropolitan nonurban	6.2	5.7	11.0
Postneonatal[a]			
All urban[b]	3.9	3.3	6.2
Metropolitan urban	3.8	3.2	6.2
Nonmetropolitan urban	4.1	3.7	6.1
All nonurban	3.4	3.1	6.1
Metropolitan nonurban	3.1	2.9	5.5
Nonmetropolitan nonurban	3.9	3.4	6.5

Source: National Center for Health Statistics, Vital Statistics of the United States, 1985, 1986, 1987, Vol. II, Mortality, Part B (Washington, DC: Government Printing Office, 1988–89), Sec. 8, Table 2.

[a]Deaths per 1,000 live births in specified group.

[b]Places with 10,000 or more population.

places for black infants, but whites are in greatest jeopardy in the small cities that lie well away from those huge centers. Conversely, whites enjoy the lowest infant mortality rates in the nonurban parts of metropolitan counties—mostly the suburbs of large cities. Blacks have the lowest rates in the urban parts of nonmetropolitan places—the small cities—though they also fare relatively well in the suburban sections of metropolitan counties. Finally, residence in an extremely rural area often means relatively high infant mortality rates for both races, owing largely to insufficient health-care facilities in sparsely populated local areas. This is why many of the highest rates in Figure 3–1 are in numerous rural counties throughout the South and in some areas of the Midwest.[54]

Among whites, both neonatal and postneonatal mortality rates follow the general pattern for infant mortality: relatively high rates in the metropolitan urban population and fairly low ones in the metropolitan nonurban population. Among blacks, neonatal mortality follows the general infant mortality pattern: high rates in metropolitan core areas and lower ones in nonmetropolitan urban places. Black postneonatal mortality, however, is highest in the least urbanized parts of the country and lowest in the suburbs of MSAs. This reflects problems in providing care for infants who have survived the first 27 days, and while inner cities present various hazards, rural areas are even riskier. Rural rates undoubtedly would be even higher if it were not for the contrast between rural and urban family structure. The proportion of mother-only families significantly affects the infant mortality rate, and in 1988, 17 percent of all nonfarm families but only 4 percent of the farm units were headed by a woman with no husband present. As a result, 96 percent of all farm children were living with both parents, compared with only 78 percent of all nonfarm children.[55]

CAUSES OF INFANT MORTALITY

Causes of infant mortality fall into four large categories: (1) *congenital anomalies*, which are mental or physical conditions present at birth that are either hereditary or the result of some condition during pregnancy; (2) *sudden infant death syndrome* (SIDS), which is a fairly recently defined respiratory syndrome; (3) *certain conditions of the perinatal period*, which may be related to pregnancy, labor and delivery, slow growth of the fetus, or birth trauma; and (4) *preventable causes*, which include infectious diseases, accidents, and homicide.[56] These large groupings account for over 90 percent of all infant deaths.[57]

Infant death can also be attributable to somewhat more specific causes under the large groupings. Moreover, death often has several contributing conditions even among infants, and single causes or even categories of causes frequently do not account for those conditions.[58]

Cause of Death and Race

Table 3–10 shows infant mortality rates for the leading official causes of death in 1985–87, separately for blacks and whites, with the black/white ratio for each cause. However, 23 percent of all white deaths and 28 percent of all black deaths fall into a residual category because many infants, especially blacks, still die from poorly defined causes. Cause of death is also reported more ambiguously among black than white infants, sometimes because black deaths have more contributing causes.

Congenital anomalies lead the list by a wide margin for both races. They include defects of the circulatory system, respiratory system, genitourinary system, and musculoskeletal system; Down's syndrome and other chromosomal defects; and anomalies of the central nervous system. Unlike other causes of death, however, congenital anomalies produce a small black/white mortality differential and the ratio is only 105. No such similarity exists for any other

Table 3–10

Infant Mortality Rate for Leading Causes of Death, by Race, 1985–87

Cause	All Races[a]	White[a]	Black[a]	Black/White Ratio
All causes	1,036.0	896.1	1,803.0	201.2
Congenital anomalies	218.1	218.8	230.6	105.4
Sudden infant death syndrome	139.7	123.0	227.4	184.9
Other respiratory conditions of newborn	96.0	79.4	186.8	235.3
Respiratory distress syndrome	91.7	82.8	146.5	176.9
Disorders relating to short gestation and low birth weight	87.0	60.7	225.5	371.5
Newborn affected by maternal complications of pregnancy	36.1	31.1	65.2	209.6
Intrauterine hypoxia and birth asphyxia	25.9	21.7	49.2	226.7
Accidents and adverse effects	24.3	21.0	40.9	194.8
Infections specific to perinatal period	24.1	20.4	44.5	218.1
Newborn affected by complications of placenta, cord, and membranes	22.7	20.3	36.9	181.8
Pneumonia and influenza	18.0	14.2	36.8	259.2
All other causes	252.4	202.7	512.7	252.9

Source: National Center for Health Statistics, Vital Statistics of the United States, 1985, 1986, 1987, Vol. II, Mortality, Part A (Washington, DC: Government Printing Office, 1988-90), Sec. 2, Tables 5, 6.

[a]Deaths per 100,000 live births in specified group.

cause, and the black/white ratio ranges from 177 to 372. Controlling for mother's education and age, the timing of prenatal care, and the infant's birth weight reduces these racial differentials, but they do not disappear. Thus, there may be racial differences in the degree of maturity at the same birth weight, though this and other possible biological variations remain to be investigated.

SIDS, or "crib death," is recorded as the second leading cause of death. It remains rather mysterious, and while some SIDS fatalities are due to an identifiable respiratory syndrome, many actually happen precipitously without a clearly identifiable cause. Perhaps as many as 90 percent result from infections or other preventable causes.[59] In fact, in the official list of causes, SIDS falls under a broader heading of "symptoms, signs, and ill-defined conditions." Therefore, after sudden death has been identified, even the most likely cause is often not recorded and considerable ambiguity persists.[60] Among whites the infant mortality rates for SIDS is a little over half that for congenital anomalies, but among blacks SIDS ranks at almost the same level as congenital anomalies. This further substantiates the tendency for causes among black infants to be ill-defined and attributed to general categories.

Respiratory distress syndrome ranks third among white infants, but disorders relating to short gestation and low birth weight are third among black infants, fifth among whites. In addition, blacks are 3.7 times as likely as whites to die from those disorders. This is consistent with earlier observations about birth weight variations between the races and the conditions that accompany low birth weight. In fact, the attack on black infant mortality is focusing increasingly on low birth weight as a major culprit. The reasons for low birth weight fall into the following five categories of risk factors carried by mothers, but to which fathers also may contribute.[61]

1. *Demographic risks* include being black, of low socioeconomic status, unmarried, and poorly educated.

2. *Medical risks* fall into two subcategories. Risks the mother carries prior to pregnancy include no pregnancies or more than three, being underweight, diseases such as diabetes and chronic high blood pressure, no immunization against certain illnesses (e.g., rubella), a poor obstetric history (e.g., an earlier low-weight birth), and maternal genetic factors (e.g., the woman's own low weight at birth). Medical risks related to a specific pregnancy include having twins or other multiple births, low weight gain, a brief interval since the last pregnancy, low or high blood pressure, certain infectious illnesses, anemia, fetal abnormalities, and various complications of pregnancy.

3. *Behavioral/environmental risks* include smoking, poor nutrition, substance abuse, and exposure to toxins (e.g., on the job), not just among mothers, but fathers as well.

4. *Health care risks* include either no prenatal care or care that begins late in the pregnancy, and infrequent visits to a health-care professional. If prenatal care begins only after 28 weeks gestation or there are fewer than five visits, it is inadequate.

5. *Evolving concepts of risk* are hazards that occur as the pregnancy progresses and that include various physical and psychological risks, certain infections, and any events that might trigger premature contractions.

Clearly, low birth weight has many contributing factors, virtually all more common among blacks than whites and which help account for the danger that low birth weight poses for infants. Even with controls for marital status, maternal age, education, income, and certain other characteristics, black women still have an unusually high incidence of low-birth-weight infants, especially if the mother herself was a low-birth-weight infant born under adverse socioeconomic conditions that affected her ability to bear an infant of normal weight.[62] This helps explain why a black woman not born in the city where she lives but who migrated from disadvantaged conditions in the South also has a high risk of an underweight birth.[63] Whatever the specifics, low birth weight and immaturity (short gestation) are major indirect causes of infant mortality and they exact a much heavier toll of blacks than whites.

Other causes of death in Table 3–10 are in almost the same descending order for blacks and whites, with high black/white mortality ratios in each case. Accidents and adverse effects, however, rank seventh for whites and eighth for blacks, with infections during the perinatal period seventh for blacks. In addition, in the list of 61 causes reported by the NCHS, the infant mortality rate of blacks is higher than that of whites for all except meningococcal infection, malignancies, benign tumors, and cystic fibrosis. Together, however, those illnesses account for only 0.2 percent of all infant deaths among blacks and less than 0.1 percent among whites. Blacks are less likely than whites to die of certain congenital anomalies, including anencephalus, spina bifida, anomalies of the genitourinary system, and a category of "other chromosomal anomalies." But those causes do not kill many infants either, so for all causes that do have a significant impact on infants of both races, blacks suffer considerably more than whites. In fact, relatively few causes are responsible for a large share of the deaths: in 1985–87 the first five in Table 3–10 caused 63 percent of the deaths among white infants and 56 percent among blacks.

Homicide, while less important statistically than the causes in Table 3–10, is responsible for a comparatively large share of infant deaths for a developed nation, especially among blacks. In 1985–87, the infant mortality rate for homicide was 4.9 among whites and 15.8 among blacks. Child battering and other maltreatment accounted for 33 percent of the white homicide deaths and 40 percent of the black ones; battering is especially likely if the child was unwanted and if one or both parents suffered child abuse.

Neonatal and Postneonatal Causes

The heavy concentration of infant deaths in the neonatal period reflects the relative importance of certain causes among neonatals and postneonatals and the effects of several demographic variables that influence the causes. Newborns often succumb to "congenital problems during pregnancy [that are] strongly associated with biological factors (e.g., maternal age, birth order); later deaths are more often caused by diseases and accidents and are more strongly associated with social and environmental factors (e.g., education, marital status)."[64] Thus, the five leading causes of neonatal mortality are (1) congenital anomalies, (2) disorders relating to short gestation and low birth weight, (3) respiratory distress syndrome, (4) maternal complications of pregnancy, and (5) complications of the placenta, cord, and membranes. Most of these are "endogenous" causes.[65] Four of the five leading causes among postneonatals are different; the list is (1) SIDS, (2) congenital anomalies, (3) accidents and adverse effects, (4) pneumonia and influenza, and (5) septicemia. These are largely "exogenous" causes.[66] Other causes among neonatals are infections of the perinatal period, intrauterine hypoxia and birth asphyxia, SIDS, neonatal hemorrhage, birth trauma, pneumonia and influenza, effects of maternal complications unrelated to the present pregnancy, complications of labor and delivery, accidents and adverse effects, and homicide. Other causes among postneonatals are homicide, respiratory distress syndrome, meningitis, bronchitis and bronchiolitis, cancer, gastritis and related illnesses, viral diseases, meningococcal infection, lingering effects of intrauterine hypoxia and birth asphyxia, and hernia and intestinal obstruction.

Several causes specific to newborns are no longer risk factors for postneonatals (e.g., complications of the placenta, cord, and membrane), although short gestation and low birth weight are the top two causes for both groups. Beyond the lingering effects of those conditions, postneonatals run risks that newborns generally do not yet incur, and postneonatals are more likely to succumb to contagious illnesses (e.g., meningitis), those that take time to develop (e.g., cancer), and violence (e.g., accidents, homicide). They are also much more likely to die of SIDS, although other diseases may cause many deaths attributed to it. In short, the two groups of infants share some of the same causes of death, but there are also many differences, not only in particular causes, but in the relative risk from causes that affect both groups. Even though over 60 percent of infant deaths occur to neonatals, postneonatals are 13 times more likely to die from SIDS, ten times more apt to succumb to accidents, eight times more likely to be homicide victims, and three times more likely to die from pneumonia and influenza.

TRENDS

Infant mortality trends reflect the progress a society has made in controlling causes of death. As people learn to contain certain illnesses, prevent most epidemics, and care for newborns and their mothers, the infant mortality rate decreases significantly. It has done so in the United States since 1915, when data on infant mortality were first collected, although fragmentary materials were reported as early as 1850. The reduction reflects improvements but also persistent differences according to socioeconomic level in various groups. The principal trends are as follows.

Long-Term Decrease

Reductions in the infant mortality rate have been spectacular, from 95.7 in 1915–19 to less than 10.0 at the end of the 1980s and 9.1 in 1990. Impressive as the decline is, however, the United States has not done as well as most other developed nations, and South Africa is the only industrialized country with a higher infant mortality rate, owing to conditions among its black majority. The U.S. infant mortality rate jumped temporarily because of the devastating 1917–18 influenza epidemic and it also rose modestly in some other years, but those occasions were more common for blacks than whites. The rate dropped sharply right after World War II, leveled out until the mid-1960s, and fell rapidly once more until the mid-1970s[67] (see Figure 3–2). Since then, however, progress has been relatively slow, partly because risks have increased for many poor infants and partly because significant improvements in the survival rates of neonatals have not been matched by improvements among postneonatals (see Figure 3–3). In fact, while the neonatal mortality rate fell 60 percent among whites and 49 percent among blacks between 1970 and 1987, the postneonatal mortality rate dropped only 23 percent among whites and 37 percent among blacks. Especially in the 1980s progress for postneonatals of both races was marginal at best.

This was not always so, and at various times the mortality rate fell faster for postneonatals than neonatals. Between 1960 and 1970, for example, the neonatal rate among blacks dropped only 18 percent while the postneonatal rate fell 41 percent. Among whites the decreases were 20 percent and 30 percent, respectively. These were years of intense spending on programs to improve the quality of life, reduce poverty and malnutrition, and otherwise provide a safer environment for older infants and their mothers. In the 1980s, however, medical innovations were geared more toward saving premature and severely underweight infants, so mortality rates among postneonatals, especially whites, tended to level out, while those among neonatals continued to decline. In one sense this is a defensible focus of medical efforts, since nearly two-thirds of all infant deaths strike neonatals, but it leaves older infants less well protected. In 1970, for example, 25 percent of all infant deaths were to postneonatals, but by 1988 the figure had increased to 36 percent. Since the first year of life is more hazardous than any succeeding one until age 65, both groups of infants need additional attention.

One reason why the infant mortality rates fell in the past three decades or so is the wider availability of legal abortion, whether in increasing numbers of states prior to the *Roe v. Wade* decision of the Supreme Court in 1973, or the nation as a whole after that landmark decision. Abortion is especially instrumental in eliminating fetuses at particularly high risk of death as newborns—those of teenagers and women over 35. Consequently, many high-risk fetuses who might have been born, only to die in their first few hours or days and to be registered as neonatal deaths, instead were eliminated by abortion. This accounts for some of the apparent "progress" in lowering the infant mortality rate in recent decades.

Given the international ranking of the United States on infant mortality, far more is needed to save postneonatal infants, even while neonatals benefit from some of the nation's most sophisticated and expensive technology, such as intensive-care units for the newborn. In addition, more can be done to reduce the percentage of underweight and premature births, since the cost of saving those who are extremely premature can run as high as $10,000 a day. In many cases, even if the infant survives it is left with lifetime health problems, such as defects of the heart and other organs, blindness, cerebral palsy, and mental

Figure 3–2
Infant Mortality Rate, by Race, 1940–88

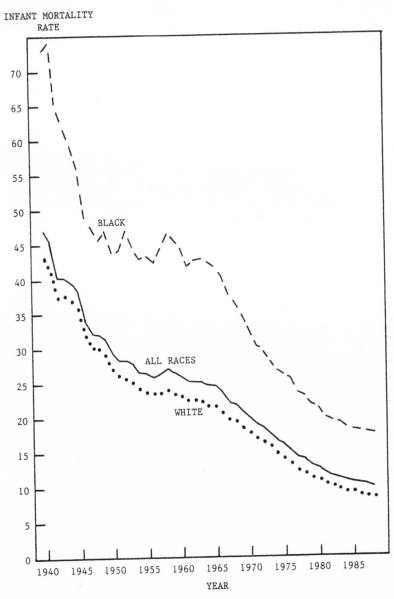

INFANT MORTALITY
RATE

BLACK

ALL RACES

WHITE

YEAR

Source: National Center for Health Statistics, *Vital Statistics of the United States, 1987*, Vol. II, *Mortality*, Part A (Washington, DC: Government Printing Office, 1990), Sec. 2, Table 1; NCHS, "Annual Summary of Births, Marriages, Divorces, and Deaths: United States, 1989," *Monthly Vital Statistics Report* 38:13 (Washington, DC: Government Printing Office, August 1990), Table 11.

Figure 3–3
Neonatal and Postneonatal Mortality Rates, by Race, 1940–88

MORTALITY RATE

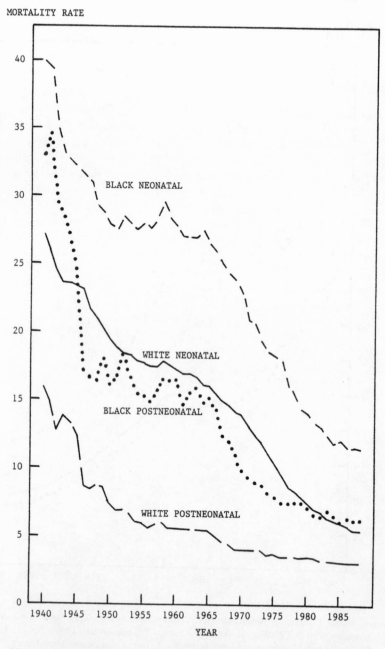

Source: National Center for Health Statistics, *Vital Statistics of the United States, 1987*, Vol. II, *Mortality*, Part A (Washington, DC: Government Printing Office, 1990), Sec. 2, Table 1.

retardation. In turn, many moral questions arise from the ability to save extremely premature infants and thereby to preserve at least a kind of life: Should we save every life because it is technologically possible, or should we apply selection criteria? If so, which criteria, identified by whom? Can we, at the birth of a defective premature infant, project its eventual quality of life?

Persistent Racial Differences

The basic relationship between blacks and whites has persisted since 1915. Decade after decade infant mortality rates among blacks have been much higher than among whites, as have the neonatal and postneonatal components (see Figures 3–2 and 3–3). There have been periodic convergences and divergences, however, as the situation of blacks relative to whites improved or deteriorated. To show the infant mortality relationship of the races between 1940 and 1988, Figure 3–4 gives black rates as a percentage of white rates for infant mortality and its neonatal and postneonatal components.

In the mid- and late 1940s, the racial difference in infant mortality was smaller than it became later because of the situation for both neonatals and postneonatals. In the few years after World War II neonatal mortality rates among blacks were about a third again as high as those among whites, but they rose to about half again as high in the early 1950s, to two-thirds greater in the late 1950s to the early 1970s, and to 80, 90, and even 100 percent higher in the late 1970s and the 1980s. This is because between 1978 and 1988 the white neonatal mortality rate fell 36 percent and the black one only 26 percent. The death controls that saved an increasing share of neonatal lives benefited whites more than blacks, partly because the costs of medical expertise are prohibitive for a larger share of blacks and partly because government funding either grew slowly or was cut in programs for the poor in their childbearing years.

After 1945 the neonatal mortality rate fell dramatically for blacks who migrated out of the poverty-stricken, segregated rural South and into northern and western cities with better jobs and other amenities that improved income, health, and other conditions related to infant mortality. By the 1980s, however, black teen-agers and young adults in many of those same cities faced high rates of unemployment, ubiquitous drugs, the exodus of middle-class role models for maternal and infant care, and other changes that slowed the decline in their neonatal mortality. Rapidly growing numbers of women also have AIDS and, therefore, the number of infants born with the disease or who test positive for HIV is also increasing, particularly among inner-city blacks, although many of these infants actually turn out not to be infected.[68]

The black/white mortality ratio for postneonatals long was much higher than for neonatals, though by the mid-1980s little difference remained and by 1988 the relationship had reversed. In all but seven years after 1940 the postneonatal mortality rate among blacks was at least twice that among whites. The greatest disparities of all were in the late 1950s and most of the 1960s, although even

Figure 3–4
Black/White Infant, Neonatal, and Postneonatal Mortality Ratios, 1940–87

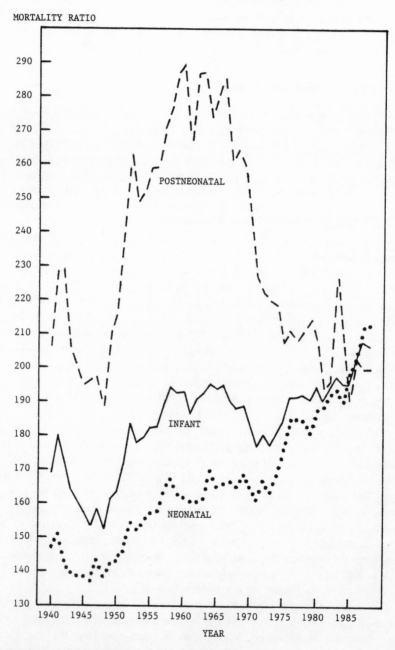

MORTALITY RATIO

YEAR

Source: National Center for Health Statistics, *Vital Statistics of the United States, 1987*, Vol. II, *Mortality*, Part A (Washington, DC: Government Printing Office, 1990), Sec. 2, Table 1.

in the best year (1948) the black ratio was nearly 90 percent higher than the white one. The postneonatal period is when various infections (e.g., influenza), accidents, homicide, certain persistent congenital anomalies, inadequate nutrition, and other causes are far better prevented and/or treated among white than black babies. Thus, until 1985 or so, the chance for a black newborn to survive was closer to that for a white newborn than was the case among postneonatals. Now, however, the difference is greater for neonatals than postneonatals, although the mortality rate among blacks in both categories is still twice that among whites. These are some of the reasons that the Surgeon General's 1990 goal of an infant mortality rate of 9.0 or below was met for whites but not blacks. That disparity will continue as long as overall social inequality persists between the races; medical and technological improvements cannot eliminate the racial differential as long as they are less accessible to blacks than whites.[69]

Static Low Birth Weight and Prematurity

A high incidence of low birth weight is "a stronger predictor of cause-specific infant mortality than [several] other independent variables combined," those being race and ethnicity, the mother's marital status, and any earlier fetal or infant deaths.[70] The effects are especially apparent in the death of neonatals and in the probability of developing long-term disabilities. In fact, about 60 percent of all infants who die are of low birth weight and about 40 percent of those under 2,500 grams are of very low birth weight (less than 1,500 grams).[71]

In 1979 what was then the Department of Health, Education, and Welfare articulated a set of health and mortality goals to be achieved by 1990. Those goals, focused on the five major stages of life and set forth in *Healthy People: The Surgeon General's Report on Health Promotion and Disease Prevention*, were realistic and potentially attainable for the United States as a whole, given its wealth, expertise, and technological sophistication. One of those goals was no more than 9.0 percent of newborns weighing under 2,500 grams. What kind of progress did we make toward that end? Unfortunately, from 1970 to 1988 we made very little, as shown in Figure 3–5. There was some improvement for whites from the early to the late 1970s, but virtually none in the 1980s: at the beginning and end of that decade just under 6 percent of all white births were low weight. Moreover, there was no change at all in the proportion under 1,500 grams; it remained at 0.9 percent. There was some minor improvement for blacks because the percentage of underweight births fell in the 1970s, but not in the 1980s. The percentage of very low weight black infants remained virtually unchanged in the 1970s and even rose some in the 1980s, to almost 3 percent in 1988. Thus, compared with the base year (1970), no improvement in any year brought the proportion of underweight births down significantly for either race.

The incidence of very low birth weight never varied for whites and its greatest drop for blacks was only 8 percent. Most brief declines were followed by in-

Figure 3–5
Percent of Births Less Than 1,500 Grams and Less Than 2,500 Grams, by Race, 1970–88

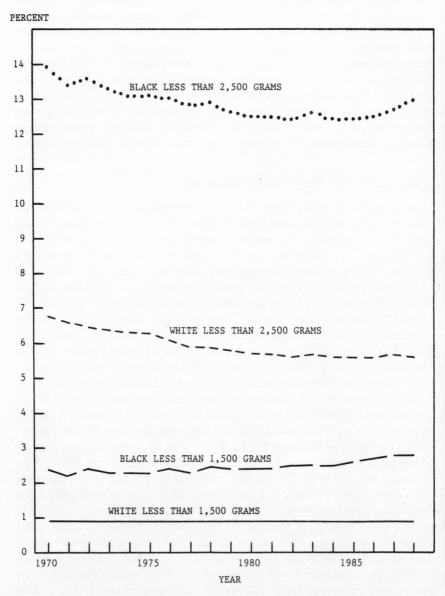

PERCENT

BLACK LESS THAN 2,500 GRAMS

WHITE LESS THAN 2,500 GRAMS

BLACK LESS THAN 1,500 GRAMS

WHITE LESS THAN 1,500 GRAMS

YEAR

Source: National Center for Health Statistics, *Vital Statistics of the United States*, 1970–88, Vol. I, *Natality* (Washington, DC: Government Printing Office, 1975–90), Sec. 1, Table 39; NCHS, "Advance Report of Final Natality Statistics, 1988," *Monthly Vital Statistics Report* 39:4 (Washington, DC: Government Printing Office, August 1990), Table 15.

creases, however, and the principal trend in low birth weight and very low birth weight is their persistence. Despite the disappointing changes, the percentages of 2,500-gram and 1,500-gram infants may seem fairly low, but in 1988 they translated into over a quarter million births of infants under 2,500 grams, almost 47,000 of them under 1,500 grams. These babies account for a large percentage of infant deaths. The ones who live often deplete their parents financially and emotionally or require high public expenditures per infant; they have very high risks of lifetime health problems and of producing underweight infants of their own with the same serious problems.

The gap between the races in the probability of low birth weight also persists. At no time since 1970 did the percentage of underweight births among blacks fall below twice that among whites. In 1970–73 the black/white ratio was 207 and by 1986–88 it had climbed to 226. The ratio for very low birth weight was 259 in 1970–73 and 307 in 1986–88. These ratios are the result of very little improvement in the underweight birth problem for either race and some increase in the relative disadvantage of blacks. Both conditions help keep the infant mortality rate, especially the neonatal rate, significantly higher than those in nearly two dozen other developed nations, although the situation is not quite so unfavorable if stillbirth rates are included in the comparisons.[72]

Prematurity, or birth before 37 weeks gestation, is closely tied to the problem of low birth weight and also contributes to infant mortality. Unfortunately, the data before 1981 are not very useful, since "length of gestation could not be determined from a substantial number of live birth certificates each year because the day of LMP [last menstrual period] was missing."[73] Even so, from 1981 to 1988 there was no decrease—even some minor increase—in prematurity among both races. Among whites the incidence of prematurity rose from 7.9 percent of all births in 1981 to 8.2 percent in 1988; among blacks it increased from 17.0 to 17.3 percent. Neither race experienced any significant reductions in the interim. As in the case of low birth weight, prematurity often results from inadequate prenatal care and maternal smoking and drug and alcohol use; from very young or relatively old age at delivery; and from socioeconomic disadvantages. Substance abuse is particularly dangerous, and about 400,000 babies are born annually to mothers who are substance abusers especially likely to produce premature infants.[74] As yet, we cannot quantify the father's contribution to this problem. Thus, we see complex interactions between (1) low birth weight and prematurity; (2) prenatal factors and conditions that cause them; (3) proximate causes of infant death and disability; (4) life chances of the underweight and premature infants who do survive; and (5) psychosocial and economic consequences for parents and the larger society. It is imperative in the study of infant mortality to understand these complex interactions in order to lower the infant mortality rate.[75]

More Births to Unmarried Mothers

The changing proportion of births to unmarried women also affects this interaction, since infant mortality rates from delivery complications and congenital anomalies are higher among unmarried than married mothers, although the problems are worse at higher than lower birth weights among the unmarried.[76] Furthermore, while older white single women have higher infant mortality rates than older white married women, the difference is minimal for blacks and teenagers, as discussed earlier. To the extent unmarried motherhood and its concomitants do involve risks to infants, the increase in unmarried childbearing contributes to the nation's relatively high infant mortality rate. Thus, the percentage of unmarried mothers rose dramatically in the 1970s and 1980s, though it actually increased fairly steadily since 1940. In that year, for example, unmarried women produced only 1.8 percent of all white births and 13.7 percent of all black ones, but by 1970 the figures had risen to 5.7 percent among whites and 37.6 percent among blacks, and by 1988 they had soared to 17.7 and 63.5 percent, respectively.

Persistent Sex Differentials

The ratio of male to female infant deaths has changed very little among both blacks and whites. From 1970 to 1988 the sex ratio showed the expected fluctuations because of reporting adequacy and other year-to-year influences. There is no consistent pattern of improvement or deterioration in the mortality relationship between male and female infants, however, and white males continue to die at rates that are a quarter to a third higher than those of white females (see Table 3–11). Black males die at rates that are 14 to 25 percent higher than rates for black females. These patterns appear in both neonatal and postneonatal mortality. There was a slight tendency for the sex mortality ratio to decrease among white neonatals and to rise among white postneonatals; among black neonatals it fell somewhat until the early 1980s and then rose slightly, while among black postneonatals the only changes were minor annual fluctuations. Thus, while infant mortality continues to drop for males and females of both races, there is no differential pace of decline sufficient to alter the gap between the sexes.

This relatively static situation results from some causes of death that take an especially high toll of males and others that differ little by sex, though the infant mortality rate among males remains higher than that among females for each major cause. In the white population the overall infant sex mortality ratio is 127, but it is even greater for respiratory distress syndrome, SIDS, infections specific to the perinatal period, homicide, and other respiratory conditions of the newborn. For the other causes in Table 3–10 and lesser ones not on the list, the sex gap in infant mortality is below average, but it persists. In the black population the overall infant mortality sex ratio is 123, but the gap is even larger for respiratory distress syndrome; complications of the placenta, cord, and membranes; other respiratory conditions of the newborn; maternal complications of pregnancy; and infections specific to the perinatal period. The sex differences for pneumonia

Table 3–11
Sex Ratio for Infant, Neonatal, and Postneonatal Mortality, 1970–88

Year	White			Black		
	Infant	Neonatal	Postneonatal	Infant	Neonatal	Postneonatal
1970	129.9	130.3	125.7	124.8	126.4	121.3
1971	131.3	132.1	128.6	122.0	125.9	114.9
1972	132.9	134.3	128.6	121.6	123.3	117.9
1973	131.6	133.7	129.4	122.6	124.4	120.0
1974	131.3	131.3	124.2	122.0	123.4	121.9
1975	129.3	130.0	127.3	118.0	119.0	115.9
1976	126.5	125.9	128.1	119.8	119.6	121.7
1977	129.9	128.9	132.3	121.6	121.8	122.0
1978	130.1	129.2	133.3	120.4	120.3	122.4
1979	129.3	127.5	133.3	119.7	118.3	122.4
1980	128.1	125.8	133.3	120.1	119.5	121.2
1981	127.2	123.8	137.9	118.6	118.7	118.3
1982	125.8	125.0	134.5	121.5	121.2	122.0
1983	125.6	124.6	127.6	122.7	121.4	125.0
1984	126.5	123.6	132.1	117.2	116.5	118.3
1985	132.5	130.2	137.0	120.6	121.1	119.6
1986	128.2	125.5	133.3	125.0	123.8	127.3
1987	126.3	124.5	129.6	122.5	122.9	121.8
1988	128.4	122.9	133.3	118.0	120.2	113.8

Source: National Center for Health Statistics, Vital Statistics of the United States, 1970–87, Vol. II, Mortality, Part A (Washington, DC: Government Printing Office, 1974–90), various tables; NCHS, "Advance Report of Final Mortality Statistics, 1988," Monthly Vital Statistics Report 39:7 (Washington, DC: Government Printing Office, November 1990), Table 13.

and influenza and accidents are also slightly above average, but the infant mortality rate for homicide is almost the same for males and females.

Respiratory problems are strongly implicated as a cause of death among both white and black male infants, and often they could be prevented or cured with appropriate care; perhaps additional research on SIDS will identify specific causes that can be ameliorated. White males are more likely than white females to be victimized by violence, and efforts to reduce child abuse and the drug, alcohol, and other problems that contribute to it should reduce that sex gap. Black male infants are more likely than females to die of problems associated with pregnancy and birth, and better prenatal and infant care could narrow that sex gap. In short, part of the sex differential is due to females' superior biological potential for survival, but part is also due to causes that could be prevented.

Changes in Cause of Death

The relative importance of various causes of infant deaths has changed significantly. In general, infectious and parasitic diseases are much better controlled, leaving such causes as congenital anomalies and others shown in Table 3–10 responsible for a larger share of infant deaths, even while the mortality rates for those causes have dropped. Pneumonia and influenza, for example, caused 16 percent of the infant deaths in 1940 but less than 2 percent in 1987. Whooping cough killed over 2,000 infants in the former year but none in the latter. Conversely, congenital anomalies caused 10 percent of the deaths in 1940 and ranked third on the list of causes, but in 1987 they caused 21 percent and ranked first, even while the infant mortality rate for that group of causes fell from 468 to 207 per 100,000 live births. In 1940, premature birth was the leading cause of infant deaths, accounting for 29 percent, but while short gestation and low birth weight remain serious problems, by 1987 they caused only 9 percent of the deaths and had fallen to fifth place.

Changes in the relative importance of various causes reflect substantial improvements in prenatal care, better conditions at delivery, more sophisticated lifesaving techniques for premature and underweight infants, and a reduction in many risks to mothers and infants. Some improvements are difficult to gauge because of periodic revisions in the terminology and categories of the *International Classification of Diseases*, but all trends indicate substantial progress. Some of it continues at a much slower rate, so there is room for even more attention to infant mortality, especially among teenage and disadvantaged mothers. Among some groups, however, such as affluent Americans of many races, infant mortality may already be close to an irreducible minimum.

NOTES

1. Regina McNamara, "Infant and Child Mortality," in *International Encyclopedia of Population*, Vol. 1, edited by John A. Ross (New York: Free Press, 1982), p. 342. See also Samuel H. Preston, ed., *The Effects of Infant and Child Mortality on Fertility* (New York: Academic Press, 1978).

2. McNamara, "Infant and Child Mortality," pp. 341–342.

3. Population Reference Bureau, *World Population Data Sheet*, 1988, 1989, 1990 (Washington, DC: Population Reference Bureau, 1988–90).

4. Lester R. Brown and Jodi L. Jacobson, *Our Demographically Divided World*, Worldwatch Paper 74 (Washington, DC: Worldwatch Institute, 1986), p. 7.

5. Population Reference Bureau, *Data Sheet*.

6. The total fertility rate is an estimate of the average number of children a woman will have if the present age-specific birth rates remain constant during her childbearing years. See John Saunders, *Basic Demographic Measures* (Lanham, MD: University Press of America, 1988), pp. 32–33.

7. Isaac W. Eberstein, "Demographic Research on Infant Mortality," *Sociological Forum* 4:3 (September 1989), p. 412.

8. John R. Weeks, "The Demography of Islamic Nations," *Population Bulletin* 43:4 (Washington, DC: Population Reference Bureau, 1988), p. 29.

9. Ibid., p. 30.

10. Robert Repetto, "Population, Resources, Environment: An Uncertain Future," *Population Bulletin* 42:2 (Washington, DC: Population Reference Bureau, 1987), p. 3.

11. For a technique to compare risks of infant death in populations with compatible data and definitions, see Carl Haub and Machiko Yanagishita, "Infant Mortality: Who's Number One?" *Population Today* 19:30 (March 1991), pp. 6–8.

12. Judith Senderowitz and John M. Paxman, "Adolescent Fertility: Worldwide Concerns," *Population Bulletin* 40:2 (Washington, DC: Population Reference Bureau, 1985), p. 3.

13. Ibid., p. 21.

14. Ibid., pp. 23–24.

15. National Center for Health Statistics, *Health, United States, 1989* (Washington, DC: Government Printing Office, 1990), p. 12.

16. Ibid., pp. 12–13.

17. National Center for Health Statistics, *Vital Statistics of the United States, 1987*, Vol. I, *Natality* (Washington, DC: Government Printing Office, 1989), Sec. 4, p. 9.

18. John R. Weeks, *Population: An Introduction to Concepts and Issues*, 4th ed. (Belmont, CA: Wadsworth, 1989), p. 171.

19. Greg J. Duncan and Saul D. Hoffman, "Welfare Benefits, Economic Opportunities, and Out-of-Wedlock Births Among Black Teenage Girls," *Demography* 27:4 (November 1990), p. 532.

20. Sandra Blakeslee, "Father Figures: The Male Link to Birth Defects," *American Health* 10:3 (April 1991), p. 56.

21. NCHS, *Natality*, Sec. 4, p. 9.

22. For a study of this matter, see Andrew Friede, Wendy Baldwin, Philip H. Rhodes, James W. Buehler, Lilo T. Strauss, Jack C. Smith, and Carol J. R. Hogue, "Young Maternal Age and Infant Mortality: The Role of Low Birth Weight," *Public Health Reports* 102:2 (March-April 1987), pp. 192–199.

23. Christiane B. Hale, *Infant Mortality: An American Tragedy*, occasional paper 13 in the series Population Trends and Public Policy (Washington, DC: Population Reference Bureau, 1988), p. 1. See also Marie C. McCormick, "The Contribution of Low Birth Weight to Infant Mortality and Childhood Morbidity," *New England Journal of Medicine* 312:2 (January 10, 1985), p. 88.

24. Carol J. R. Hogue, James W. Buehler, Lilo T. Strauss, and Jack C. Smith, "Overview of the National Infant Mortality Surveillance (NIMS) Project—Design, Methods, Results," *Public Health Reports* 102:2 (March-April 1987), pp. 130–132.

25. Eberstein, "Research on Infant Mortality," p. 420.

26. Hale, *Infant Mortality*, p. 1.

27. Margaret S. Boone, *Capital Crime: Black Infant Mortality in America* (Newbury Park, CA: Sage, 1989), pp. 16, 25.

28. Weeks, *Population*, p. 171.

29. James C. Cramer, "Social Factors and Infant Mortality: Identifying High-Risk Groups and Proximate Causes," *Demography* 24:3 (August 1987), pp. 310, 316.

30. Ibid., p. 310.

31. Greg J. Duncan and Saul D. Hoffman, "A Reconsideration of the Economic Consequences of Marital Dissolution," *Demography* 22:4 (November 1987), p. 485.

32. Jonathan Crane, "The Epidemic Theory of Ghettos and Neighborhood Effects on Dropping Out and Teenage Childbearing," *American Journal of Sociology* 96:5 (March 1991), p. 1227.

33. For the data on poverty and family form, see U.S. Bureau of the Census, "Measuring the Effects of Benefits and Taxes on Income and Poverty: 1989," *Current Population Reports*, Series P–60, No. 169-RD (Washington, DC: Government Printing Office, 1990), Table 2.

34. For a discussion of economic improvement among rural minorities between 1959 and 1979 and subsequent reversals, see Leif Jensen and Marta Tienda, "Nonmetropolitan Families in the United States: Trends in Racial and Ethnic Stratification," *Rural Sociology* 54:4 (Winter 1989), pp. 509–532.

35. Eberstein, "Research on Infant Mortality," p. 415.

36. Isaac W. Eberstein, Charles B. Nam, and Robert A. Hummer, "Infant Mortality by Cause of Death: Main and Interaction Effects," *Demography* 27:3 (August 1990), p. 414.

37. Wendy H. Baldwin and Christine Winquist Nord, "Delayed Childbearing in the U.S.: Facts and Fictions," *Population Bulletin* 39:4 (Washington, DC: Population Reference Bureau, 1985), pp. 24–25.

38. For cautions about the use of these data and a list of the states, see National Center for Health Statistics, *Vital Statistics of the United States, 1987*, Vol. II, *Mortality*, Part A (Washington, DC: Government Printing Office, 1990), Sec. 7, p. 6.

39. Cary Davis, Carl Haub, and JoAnne Willette, "U.S. Hispanics: Changing the Face of America," *Population Bulletin* 38:3 (Washington, DC: Population Reference Bureau, 1983), p. 18.

40. Hale, *Infant Mortality*, p. 1.

41. For the economic data, see U.S. Bureau of the Census, "Money Income of Households, Families, and Persons in the United States: 1987," *Current Population Reports*, Series P–60, No. 162 (Washington, DC: Government Printing Office, 1989), Table 19; Bureau of the Census, *Statistical Abstract of the United States: 1990* (Washington, DC: Government Printing Office, 1990), Tables 43, 44, 45.

42. National Center for Health Statistics, Barbara Bloom, "Health Insurance and Medical Care: Health of Our Nation's Children, United States, 1988," *Advance Data from Vital and Health Statistics* 188 (Washington, DC: Government Printing Office, October 1990), pp. 2–5.

43. National Center for Health Statistics, Peter Ries, "Health Care Coverage by Age, Sex, Race, and Family Income: United States, 1986," *Advance Data from Vital and Health Statistics* 139 (Washington, DC: Government Printing Office, September 1987), p. 2.

44. NCHS, "Health Insurance and Medical Care," pp. 2–5.

45. Boone, *Capital Crime*, p. 58.

46. National Center for Health Statistics, William F. Pratt and Marjorie C. Horn, "Wanted and Unwanted Childbearing: United States, 1973–82," *Advance Data from Vital and Health Statistics* 108 (Washington, DC: Government Printing Office, May 1985), pp. 1, 3.

47. Boone, *Capital Crime*, p. 29.

48. For a discussion of the impact of the mother-only family on the "feminization of poverty," see Paul E. Zopf, Jr., *American Women in Poverty* (Westport, CT: Greenwood Press, 1989), pp. 13–14. See also Martha S. Hill, "The Changing Nature of Poverty,"

in "The Welfare State in America: Trends and Prospects," edited by Yeheskel Hasenfeld and Mayer N. Zald, *Annals of the American Academy of Political and Social Science* 479 (May 1985), pp. 31–47.

49. National Center for Health Statistics, *Health, United States, 1988* (Washington, DC: Government Printing Office, 1989), p. 79. See also Assistant Secretary of Health and Surgeon General, *Healthy People: The Surgeon General's Report on Health Promotion and Disease Prevention* (Washington, DC: Government Printing Office, 1979).

50. Hale, *Infant Mortality*, pp. 9–11.

51. For an account of these changes, see ibid.

52. For a study of state differentials, see James S. Marks, James W. Buehler, Lilo T. Strauss, Carol J. R. Hogue, and Jack C. Smith, "Variation in State-Specific Infant Mortality Risks," *Public Health Reports* 102:2 (March-April 1987), pp. 146–161.

53. See, for example, the cooperative studies on infant mortality carried out in Alabama by J. Selwyn Hollingsworth, in Georgia by Everett S. Lee and Douglas C. Bachtel, in Mississippi by Mohamed El-Attar, and in South Carolina by C. Jack Tucker, under the rubric "Infant Mortality in the Deep South," papers presented at the Southern Demographic Association (Durham, NC: October 18–20, 1989).

54. Hale, *Infant Mortality*, p. 9. See also David M. Allen, James W. Buehler, Carol J. R. Hogue, Lilo T. Strauss, and Jack C. Smith, "Regional Differences in Birth Weight-Specific Infant Mortality, United States, 1980," *Public Health Reports* 102:2 (March-April 1987), pp. 138–145.

55. U.S. Bureau of the Census, "Residents of Farms and Rural Areas: 1989," *Current Population Reports*, Series P-20, No. 446 (Washington, DC: Government Printing Office, 1990), p. 12.

56. Hale, *Infant Mortality*, p. 7. See also Hogue et al., "Overview."

57. Hale, *Infant Mortality*, p. 6.

58. Eberstein, "Research on Infant Mortality," p. 416.

59. Hale, *Infant Mortality*, p. 7.

60. Michael Alderson, *Mortality, Morbidity and Health Statistics* (New York: Stockton Press, 1988), p. 59.

61. Reported by Hale, *Infant Mortality*, p. 8, from Institute of Medicine, *Preventing Low Birth Weight* (Washington, DC: National Academy of Sciences Press, 1986), p. 51.

62. Hale, *Infant Mortality*, p. 8.

63. Boone, *Capital Crime*, p. 29.

64. Cramer, "Social Factors and Infant Mortality," p. 311.

65. A. H. Pollard, Farhat Yusuf, and G. N. Pollard, *Demographic Techniques*, 3d ed. (Sydney, Australia: Pergamon Press, 1990), p. 68.

66. Ibid., p. 69.

67. Hale, *Infant Mortality*, p. 2.

68. Kristen Bruno, "AIDS Update," *American Health* 10:3 (April 1991), p. 9.

69. Eberstein et al., "Infant Mortality by Cause of Death," p. 427.

70. Ibid., pp. 415, 419.

71. NCHS, *Health, 1989*, p. 12.

72. Haub and Yanagashita, "Infant Mortality," p. 6.

73. NCHS, *Natality*, Sec. 4, p. 9.

74. NCHS, *Health, 1989*, p. 12.

75. Eberstein et al., "Infant Mortality by Cause of Death," pp. 425–426.

76. Ibid., p. 426.

4 Cause of Death: Differentials and Distribution

Along with infant mortality, cause-of-death analyses are among the most important in mortality studies. The reported cause of death is a statistical concept, however, for even though the death certificate reports immediate and underlying causes and other significant conditions that contributed to a death, only one cause is used in compiling the aggregate data.[1] Yet death is often caused by *comorbidity*, or several chronic conditions that together may lead to death, especially among the elderly. In fact, about a third of all deaths are due to two causes, a quarter to three, and about 16 percent to four or more.[2]

Deaths from multiple causes can be analyzed using the *underlying cause* and contributing conditions, and numerous studies have dealt with multiple causes for specific groups whose individual death certificates are available.[3] That prospect is limited though not impossible in large populations, studies of which generally rely on the single reported cause of death. In order to compensate partially for this situation, since 1949 the NCHS has based cause-of-death statistics on the underlying cause of death, which is often different from the immediate cause. The World Health Organization, in its Ninth Revision of the *International Classification of Disease, Injuries, and Causes of Death* (1975), defines the underlying cause as "(a) the disease or injury which initiated the train of events leading directly to death, or (b) the circumstances of the accident or violence which produced the fatal injury."[4]

The single underlying cause "is a simple, one-dimensional statistic; it is conceptually easy to understand and a well-accepted measure of mortality."[5] Therefore, aggregate data used in this chapter are based on the underlying cause of death, but they reflect neither contributing causes nor even necessarily the one that finally killed a given individual. For example, a woman with terminal breast cancer, who also suffered from anemia, whose malignancy metastasized

124 • MORTALITY PATTERNS AND TRENDS IN THE UNITED STATES

to other sites, and who was so weakened in the last stages of her illness that she succumbed to pneumonia, is grouped with women for whom breast cancer is reported as the cause of death. Thus, aggregate data have significant but unavoidable limitations, although some contributing causes are mentioned in the sections that follow.

The data also need age controls, and the topics in this chapter use age-adjusted or age-specific death rates for each cause, although some conclusions are reported for large age groups (e.g., 5–14, 35–64 . . . 65 and over). The need for age controls is reflected in death rates for diseases of the heart. In 1986–88 the crude death rate for that group of causes was 314 deaths per 100,000 population, but the rates for all age groups under 35 were less than 10, whereas by the ages 65–74 the rate was over 1,000, and it climbed even higher among older persons. Since the percentage of elderly is rising, the relatively high crude rate for heart disease is strongly influenced by that aging. The age-adjusted death rate for heart disease is only 170, so the distortions diminish if we analyze cause-specific rates by age, sex, and race.

For some purposes the study of cause-specific mortality rates is less satisfactory than the study of *morbidity*, or the extent of disease, injury, and disability in a population. Data on morbidity better describe a people's well-being because mortality data do not reflect certain chronic health problems that afflict large numbers of people in an aging population, such as arthritis, hypertension, cataracts, heart disease, varicose veins, diabetes, and others.[6] Nor do the mortality data provide much information on mental illness or retardation, or on long-term suffering from certain maladies, such as cancer, alcoholism, and Alzheimer's disease, though they may eventually cause death. Therefore, while morbidity data are not as complete as those on mortality and while a comprehensive analysis of morbidity is beyond the scope of this chapter, those materials identify the illnesses and other health problems that affect a population's quality of life.

TEN MAJOR CAUSES

The ten leading causes of death, ranked according to their age-adjusted death rates among all Americans collectively, are identified below. Most are discussed more extensively in later sections of this chapter and in Chapter 5, but each merits a few comments here. The basic data appear in Table 4–1, which includes rates for blacks and whites, males and females. Causes conform to the categories in the Ninth Revision of the *International Classification of Diseases*, but for simplicity the category numbers are omitted from Table 4–1 and others.

Diseases of the Heart

Americans are significantly more likely to die from heart disease than from malignant neoplasms (cancer), the second leading cause of death, though the mortality gap between them has narrowed: in 1974 cancer caused 49 percent as

Table 4–1
Age-Adjusted Death Rate for Selected Causes, by Race and Sex, 1986–88

Cause	All Races[a]	White[a] Male	White[a] Female	Black[a] Male	Black[a] Female
All causes	537.6	670.8	385.4	1,029.3	589.2
Diseases of heart	170.3	227.1	116.5	289.2	182.3
Malignant neoplasms	132.9	158.3	110.0	228.0	131.8
Accidents and adverse effects	34.9	50.2	18.6	67.6	21.4
Cerebrovascular diseases	30.3	30.5	26.3	57.9	47.0
Chronic obstructive pulmonary diseases	19.0	27.8	13.8	24.9	9.5
Pneumonia and influenza	13.6	17.4	10.1	27.2	12.9
Suicide	11.7	20.1	5.3	11.8	2.3
Diabetes mellitus	9.8	9.4	8.2	18.7	21.6
Chronic liver disease and cirrhosis	9.1	12.1	5.2	21.2	9.2
Homicide and legal intervention	8.9	7.9	2.9	56.0	12.3
Human immunodeficiency virus infections	6.1	9.1	0.7	28.5	5.5
All other causes	91.1	100.9	67.8	198.3	133.4

Source: National Center for Health Statistics, Vital Statistics of the United States, 1986, 1987, Vol. II, Mortality, Part A (Washington, DC: Government Printing Office, 1988–90), various tables; NCHS, "Advance Report of Final Mortality Statistics, 1988," Monthly Vital Statistics Report 39:7 (Washington, DC: Government Printing Office, November 1990), Tables 12, 25.

[a]Deaths per 100,000 population in specified group.

many deaths as heart disease, whereas in 1988 it caused 63 percent as many. The convergence resulted from a 25 percent decrease in the age-adjusted death rate for heart disease between 1974 and 1988, accompanied by a 1 percent increase in the age-adjusted death rate for cancer. A rise in the death rate for respiratory cancer is particularly at fault, but so are an increase in the rate for prostate cancer and virtually no decrease in the rate for breast cancer.

Ischemic heart disease, manifested as heart attack, claims about two-thirds of all heart disease victims, especially from myocardial infarction, which is tissue necrosis due to an abrupt reduction of blood flow to the myocardium (muscle tissue of the heart). Heart failure and hypertensive heart disease are also significant killers, but hypertension is a much larger contributor to heart disease than is reflected in the actual number of people who die from hypertension or hypertensive heart disease. In fact, according to the 1986 National Mortality Followback Study, 50 percent of men 65 and over and 59 percent of the women who died from ischemic heart disease had high blood pressure, over a quarter had strokes, a quarter suffered angina, and over a fifth had diabetes.[7] Among persons aged 25–64 who died from ischemic heart disease the percentage with high blood pressure was somewhat greater than among persons 65 and over, the incidence of stroke was substantially lower, and the proportions with angina and diabetes were about the same in both age groups. Atherosclerosis is also a complicating condition: in 1986–88 that cause per se had an age-adjusted death rate of only 3.6 per 100,000, but it contributed significantly to acute myocardial infarction and other forms of ischemic heart disease, often producing arterial blockage by a thrombus (blood clot) or arterial plaque and causing a heart attack. Atherosclerosis is implicated in renal failure, cerebrovascular diseases, and others, and it exemplifies conditions that contribute to the underlying cause of death.

The designation "coronary heart disease" is used in the literature, including references cited in this book. If one uses terminology strictly in accord with the causes of death in the *ICD*, "coronary" is found in connection with "coronary atherosclerosis" (*ICD* classification 414.0), which is a subcategory under ischemic heart disease, and in connection with "coronary artery anomaly" (*ICD* 746.85), which is a congenital problem. Coronary heart disease is not listed as a cause of death, though it refers to types of heart disease in which blockage and other problems of the coronary arteries are the common denominator and heart attack is the result.[8]

Malignant Neoplasms (Cancer)

Among the deaths from cancer, those attributable to the respiratory and intrathoracic organs, especially the lung, account for the largest number, followed by cancer of the digestive organs (e.g., colon, pancreas); cancer of the genital organs (e.g., prostate, ovaries); cancer of the breast (third leading cancer among women); cancer of the urinary organs; leukemia; and cancer of the lip, oral cavity, and pharynx. Smoking is strongly implicated in cancer of the lung and other respiratory organs, of the thoracic region, and of the lip, oral cavity, and pharynx, though it plays a less direct role in some other cancers as well, and in heart disease and other cardiovascular diseases.

Accidents and Adverse Effects

Accidents and adverse effects are third on the list of causes, but they cause only 20 percent as many deaths as heart disease and 26 percent as many as cancer, so the two leading two causes are far more virulent killers than any other. Motor vehicle accidents cause somewhat more than half of all accidental deaths, and falls, poisoning of various kinds, fire, and drowning produce most of the rest.

Cerebrovascular Diseases

Cerebrovascular diseases (stroke) are fourth according to the age-adjusted death rate, third according to the crude death rate. Stroke deaths often result from occlusion of the cerebral arteries by thrombosis, embolism, or other obstructions; by intracerebral hemorrhage or subarachnoid hemorrhage; or by cerebral atherosclerosis. But those causes account for only 36 percent of all deaths in the cerebrovascular category, and well over half are attributed to acute but ill-defined conditions. The total number of persons under age 65 who die annually from cerebrovascular diseases would live a collective total of about 250,000 years longer if they were spared death from that cause.[9] In addition, cerebrovascular diseases often strike without being fatal, affecting as many as 12 persons per 1,000, and in 40 percent of the cases the diseases restrict activities. Persons who die from stroke, especially women, are somewhat more likely to be confined to an institution before death than those who die from other causes.[10] As expected, high blood pressure is a common antecedent. According to the 1988 National Mortality Followback Study, for example, 60 percent of stroke decedents had high blood pressure, compared with 32 percent of persons who died from heart attack. As with heart disease, stroke is often accompanied by other diseases, especially heart attack, angina, diabetes, various lung diseases, and sometimes cancer. Much of this association is due to the typical age of most stroke victims, but part also relates to cigarette smoking and other elements of lifestyle.[11]

Chronic Obstructive Pulmonary Diseases

Chronic obstructive pulmonary diseases rank fifth. The principal subcategories are emphysema, asthma, and bronchitis, but 70 percent of the deaths are described as "chronic airways obstruction not otherwise classified," so these several diseases make up a generalized category in which most deaths cannot be attributed to specific subcategories. It is clear, however, that a huge percentage of deaths from these causes results directly or indirectly from cigarette smoking. In fact, if smoking could be eliminated entirely, this major category of deaths would subside into insignificance.[12]

Pneumonia and Influenza

Pneumonia and influenza are sixth, though pneumonia accounts for 98 percent of the deaths in that category. Bacterial pneumonia is considerably more common than viral pneumonia among cases for which the microorganism is specified, but in 90 percent of the deaths from this cause it is unspecified.

Suicide

Suicide is the seventh leading cause, largely because of the high rate among white males. Despite well-publicized increases among teenagers and young adults, the elderly have the highest suicide rates, and what had been a 50-year decrease in that group now has become an increase. Furthermore, many actual suicides among the elderly are ascribed to other events, such as automobile accidents and failure to take crucial medicines. Elderly white men are especially vulnerable to alcoholism and depression that result in suicide, but while some suicides are impulsive without warning, others follow signals that go unrecognized. Suicides choose to die by firearms in over half the cases, followed by hanging and poisoning.

Diabetes Mellitus

Diabetes mellitus, the eighth leading cause, actually refers to a diverse group of problems with glucose intolerance in common.[13] While complications of diabetes mellitus often are not recorded, many deaths from this cause are accompanied by renal, circulatory, and other problems because diabetes either precipitates those illnesses or aggravates them if they are already present at the onset of diabetes.[14] Even so, about 75 percent of the deaths from diabetes are recorded without complications because they did not occur or were not registered. Diabetes is also one of the few causes that occur more commonly among women than men, although death rates by sex are not vastly different. It is also much more an affliction of blacks than whites, and it is more frequent among persons who are less educated and poor. It shows up more often in central cities than outlying areas and is more common in the South than any other region. It afflicts about 25 persons per 1,000 in the general population, but nearly 90 in the ages 65 and over. Furthermore, about half the population with diabetes is limited in one or more activities, compared with only 15 percent of the general population.[15] Despite its indirect contribution to mortality, diabetes is not the underlying cause of a huge number of deaths, and the 40,000 persons it killed in 1988 were less than 2 percent of all deaths.

Chronic Liver Disease and Cirrhosis

Chronic liver disease and cirrhosis are ninth, and cirrhosis accounts for 76 percent of those deaths, particularly among males because of alcohol-use patterns. Alcohol is not officially implicated, however, in about half the deaths from cirrhosis.

Homicide and Legal Intervention

Homicide and legal intervention (e.g., execution, shot by police officer) are tenth, although for black males this is the fifth leading cause of death. Among all races collectively, homicide accounts for 99.9 percent of the category, so legal intervention is statistically insignificant. About 61 percent of homicides are caused by firearms, but 69 percent of black male victims die by that means. In fact, male homicide victims of both races are much more likely to die by firearms than are female victims. Assault by cutting and piercing instruments is the next most common means, and black females are the most likely victims of those instruments. Furthermore, homicide is more common among infants and adults than among children aged 1–14, not just in the United States, but in most other developed nations as well.[16]

These ten leading causes account for about 83 percent of all deaths in the United States, even though the NCHS reports deaths for hundreds of other causes. Moreover, in the past few decades the age-adjusted death rates for heart disease, accidents and adverse effects, cerebrovascular diseases, pneumonia and influenza, diabetes mellitus, and some lesser ones have gone down significantly, but those for cancer have risen somewhat (for lung cancer the increase has been substantial), and the rates for chronic obstructive pulmonary diseases and homicide have risen markedly. The age-adjusted death rate for chronic liver disease and cirrhosis rose modestly until the early 1970s and declined modestly thereafter, while the rate for suicide remained essentially unchanged until recent rises among young adults and the elderly. The increase in cancer-caused deaths is especially important because of the numbers involved, and while that disease accounted for 15 percent of all deaths in 1950, by the end of the 1980s it caused 22 percent. Over the same period deaths attributable to heart disease fell from 37 to 32 percent. No matter what the specifics, the diseases that kill the large majority of Americans are deteriorative, or degenerative, and typical of a population with long average life expectancy and relatively effective death control, although pneumonia and influenza, accidents, suicide, and homicide also are important causes and could be prevented more effectively.

HIV/AIDS, DRUGS, ALCOHOL, ALZHEIMER'S DISEASE

Four other causes are worth noting because of their inherent seriousness and because they contribute to other causes. They are (1) human immunodeficiency virus (HIV) infections and acquired immune deficiency syndrome (AIDS), (2) drug-induced causes, (3) alcohol-induced causes, and (4) Alzheimer's disease. HIV/AIDS is listed in Table 4–1, but the others overlap partly with such causes as suicide and are not shown. None of the four is officially recorded in the top ten causes for the population as a whole.

HIV/AIDS

HIV/AIDS is not listed among the ten principal causes of death because of its relatively low rate among white females, but it is increasing rapidly as a cause, particularly among teenagers and black women, and may soon enter the list. Moreover, it does rank sixth as a cause among black males and ninth among white males and black females. We are in the early stages of analyzing this cause and the NCHS assigned HIV infections a separate category as a cause only in 1987, though we have data for most of the 1980s. The NCHS also has collected data on this cause since 1987 in the National Health Interview Survey.[17] Between

1987 and 1988 the age-adjusted death rate for HIV infections in all groups collectively rose 24 percent; among black females it increased 34 percent; among white males, 28 percent; among black males, 25 percent; and among white females 17 percent. Even accounting for improvements in reporting, these are substantial increases unmatched by those for any other cause of death.[18]

These changes have not altered the fact that HIV/AIDS deaths are much more heavily concentrated in some groups than others. In 1988 black males were the most vulnerable, for while they were 5.7 percent of the nation's total population they accounted for 25.3 percent of its AIDS deaths. White males made up 41.0 percent of the population and had 63.1 percent of the AIDS deaths. Black females accounted for 6.5 percent of the population and 6.0 percent of the AIDS deaths. White females were least likely to be infected because they accounted for 43.4 percent of the population but only 4.7 percent of the deaths from AIDS. In addition, Hispanic males were about 4.1 percent of the population and had 11.6 percent of the deaths from this cause, while Hispanic females were 4.0 percent of the population and had 1.5 percent of the deaths from AIDS. Thus, AIDS deaths continue to be heavily concentrated among males, especially in minority groups, though the disease is spreading in every category.

HIV is a complex underlying cause of death with many precipitating causes. In 1988, for example, 41 percent of the HIV deaths were attributable to AIDS, the rest to other infections (e.g., pneumonia), malignancies (e.g., Kaposi's sarcoma), diseases of the central nervous system, and other terminal results of HIV infections. Thus far HIV has been concentrated in certain age groups: males aged 25–44 accounted for 16 percent of the total white population but 67 percent of all HIV white deaths; males in those ages were 14 percent of the black population but had 63 percent of the HIV deaths among blacks. But this is not containment, and the disease is spreading rapidly among women and other age groups, including infants born with the disease. Moreover, HIV infections have been highly concentrated in the urban areas of a few states: in 1988 two-thirds of the cases were in New York, California, New Jersey, and Florida, though the District of Columbia had the highest death rate for HIV infections. This distribution is not containment either, since every state reported at least a few cases, and what had been a big-city phenomenon has moved quickly to other areas. Its present spread among teenagers is helping disperse it geographically among both sexes and all ages, particularly among marginalized groups that suffer "poverty, physical weakness, vulnerability, isolation and power-lessness."[19]

Part of the problem in dealing with HIV/AIDS is persistent misinformation. For example, the 1989 Secondary School Student Health Risk Survey shows that 54 percent of all high school students had some education in school about HIV/AIDS and that virtually all students knew that intravenous drug use and sexual intercourse were the principal modes of transmission.[20] But nearly half the students had no HIV/AIDS education and they had more sexual partners and less consistent condom use than students with such education. About 12 percent of the students surveyed also believed the birth control pill protects against HIV infections and 23 percent believed one could identify a person with AIDS by appearance.[21] Thus, while high-risk behavior is changing among many male homosexuals and at least some intravenous drug users, change is slower among adolescents.[22]

Knowledge about HIV/AIDS varies by level of education and age but not much by race when those other characteristics are held constant: the best-educated persons and young adults are most knowledgeable about the disease regardless of race.[23] In fact, a somewhat larger percentage of blacks than whites aged 10–17 received education in school about HIV/AIDS, but black adults feel they know less about them than do white adults, and among persons aged 50 and over that seems to be true.[24] Moreover, while blacks are about as likely as whites to know the principal modes of transmission, they are also more likely to believe that human contacts with little or no risk can cause transmission, although level of education, not race, most affects knowledge about HIV/AIDS.[25] Much the same is true of the Hispanic population, and the greatest difference in knowledge and attitudes is between those who completed 12 years of schooling and those who did not.[26]

Inadequate knowledge about the disease is by no means the only reason for its current spread. Many persons who are exposed to information about HIV/AIDS, especially the ways in which it is transmitted and how to protect themselves, fail to change their behavior to reduce their own susceptibility and that of others with whom they are in contact. In part, this failure results from the lingering belief that HIV/AIDS is essentially confined to homosexual/bisexual men and intravenous drug users, although recent publicity about the incidence of new cases among heterosexuals may help overcome that delusion. In addition, the vague sense of immortality among teenagers and young adults that leads to numerous types of risk-taking behavior may cause many who are informed about HIV/AIDS to ignore the information or to act on it haphazardly. Intravenous drug users often are aware at one level that there is danger in sharing needles, but beset by drug-induced apathy or the urgent need for the next injection they may be unable or unwilling to protect themselves against the intravenous mode of transmission. In short, while deficient knowledge of HIV/AIDS contributes to the epidemic, so does failure to change behavior despite adequate knowledge.

The importance of behavioral patterns is reflected in the percentages of all AIDS cases contracted by various means. In 1988, for persons aged 13 and over, 59 percent of the AIDS cases were among homosexual/bisexual males, while 23 percent were among intravenous drug users. Men who were homosexual/bisexual and also intravenous drug users accounted for another 7 percent. About 4 percent of the cases were from heterosexual contact, more than two-thirds of them with an intravenous drug user. That left only 7 percent of the cases attributable to blood transfusions, hemophilia/coagulation disorder, persons born in the Caribbean or Africa, and undetermined origins.

In the white population 75 percent of the cases occurred among males who were homosexual/bisexual, compared with 35 percent in the black population and 42 percent in the Hispanic group. On the other hand, heterosexual contact was responsible for only 2 percent of the AIDS cases among whites, 6 percent among blacks, and 5 percent among Hispanics. Strikingly, intravenous drug use caused 9 percent of the cases among whites, 42 percent among blacks, and 39 percent among Hispanics.

Thus, not only are specific behavior patterns strongly associated with the likelihood of contracting AIDS, but the relative importance of these patterns varies sharply by race and Hispanic origin. It also varies by sex. In 1988, for example, males were most likely to contract AIDS through homosexual/bisexual contact (66 percent), followed by intravenous drug use (19 percent), the combination of homosexual/bisexual contact and intravenous drug use (7 percent), blood transfusion (2 percent), and heterosexual contact (1 percent). Females were most likely to become infected by intravenous drug use (53 percent), followed by heterosexual contact (27 percent), and blood transfusion (11 percent). Among both sexes, over 70 percent of the AIDS cases generated by heterosexual contact came from intravenous drug users.[27]

Drug-Induced Causes

Drug-induced deaths are attributable to drug psychoses; drug dependence; nondependent use of drugs (excluding alcohol and tobacco); accidental poisoning by drugs, medicaments, and biologicals; assault from poisoning by drugs and medicaments; and other undetermined poisoning. The category does not include homicide, traffic accidents, and other indirect results of drug use.[28] Almost 11,000 persons died by drugs in 1988, and the rate has been rising steadily, from 3.1 deaths per 100,000 in 1979 to 4.2 percent in 1988. Among white females the death rate has fluctuated some, but it was lower in 1988 (2.7) than in 1979 (3.0). It has risen for white males and blacks of both sexes, however, with the most rapid increase among black males. Between 1979 and 1988 they incurred a 163 percent increase in the age-adjusted death rate for drug-induced deaths, compared with 59 percent for black females and 50 percent for white males. The death rate from drug-induced causes does not come close to those for heart disease, cancer, and several others, but among black women it is higher than the suicide rate and among black men the two rates are almost the same.

Alcohol-Induced Causes

Alcohol-induced deaths include those from alcoholic psychoses, alcohol dependence syndrome, nondependent abuse of alcohol, alcohol polyneuropathy, alcoholic cardiomyopathy, alcoholic gastritis, chronic liver disease and cirrhosis specified as alcoholic, excessive blood level of alcohol, and accidental poisoning by alcohol. As with drug-induced deaths, those from alcohol do not include indirect deaths such as accidents and homicide. In 1988 alcohol was directly responsible for 19,000 deaths—far more than those induced by drugs. Therefore, the age-adjusted rate for this cause was 6.8 in 1986–88. The rate was highest among black males (26.0), followed by white males (9.1), black females (7.3), and white females (2.7). Although the number of alcohol-induced deaths did not change much between 1979 and 1988, the rates fell 24 percent for black females, 23 percent for white females, 13 percent for black males, and 11 percent for white males. At least these are improvements, contrasted with the increases in drug-induced death rates among white males and blacks of both sexes, but rates of alcohol-induced deaths still place it on the list of ten leading causes among white males (surpassing homicide) and among black males and females (surpassing suicide). Moreover, alcohol is a contributing factor in a large number of deaths from other causes.

Alzheimer's Disease

Alzheimer's disease is listed as a cause of death in the *ICD*, but at this writing it is not separately reported by the NCHS in *Vital Statistics of the United States*. Instead, it is part of "mental disorders" and the subcategory "presenile dementia" when accompanied by dementia; otherwise it falls in the general category of "hereditary and degenerative diseases of the central nervous system." It is a significant cause of death, however, especially among the elderly. Alzheimer's disease probably claims at least 100,000 lives annually and about 4 million Americans are thought to have the disease, which often starts two decades or more before it causes death. Therefore, it is a major degenerative health problem and may actually rank fourth or fifth in the list of causes according to the crude death rate, although it is often difficult to diagnose, particularly in its early stages.

DIFFERENTIALS BY SEX

The age-adjusted death rates of males are higher than those of females for each cause in Table 4–1, regardless of race; in most cases the difference is substantial, though a few causes are sex-specific (e.g., prostate cancer, cervical cancer) and no comparison is possible. Heart disease is the major killer of males and females of both races; cancer is second. The ranking changes for certain other causes, however, when males and females are considered separately: suicide moves up to sixth place among white males; cerebrovascular diseases rise to third place among black and white females; homicide increases to fifth place among black males, and for them pneumonia and influenza, chronic liver diseases and cirrhosis, and diabetes also increase in relative importance; diabetes is the fourth leading cause among black women, but they are least likely to die of chronic obstructive pulmonary diseases and suicide.

In order to show sex variations systematically, Table 4–2 gives the male/female age-adjusted mortality ratio for each cause of death, including some not

Table 4–2
Male/Female Ratio of Age-Adjusted Death Rates for Selected Causes, by Race, 1986–88

Cause	All Races	White	Black
All causes	173.3	174.1	174.2
Suicide	387.8	379.2	513.0
Homicide and legal intervention	331.7	272.4	455.3
Accidents and adverse effects	273.5	271.5	315.9
Motor vehicle	257.3	252.2	330.7
Chronic liver disease and cirrhosis	228.1	232.7	230.4
Chronic obstructive pulmonary diseases	205.3	201.4	262.1
Diseases of heart	189.2	194.9	158.6
Ischemic heart disease	204.5	211.0	159.3
Pneumonia and influenza	176.0	172.3	210.9
Malignant neoplasms	146.7	143.9	173.0
Respiratory organs	253.8	243.5	347.7
Digestive organs	163.2	162.0	174.3
Cerebrovascular diseases	116.3	116.0	123.2
Diabetes mellitus	107.4	114.6	86.6
Human immunodeficiency virus infections[a]	846.2	1,300.0	518.2
All other causes	147.1	148.8	148.7

Source: National Center for Health Statistics, Vital Statistics of the United States, 1986, 1987, Vol. II, Mortality, Part A (Washington, DC: Government Printing Office, 1988-90), various tables; NCHS, Health, United States, 1989 (Washington, DC: Government Printing Office, 1990), Table 23; NCHS, "Advance Report of Final Mortality Statistics, 1988," Monthly Vital Statistics Report 39:7 (Washington, DC: Government Printing Office, November 1990), Tables 12, 25.

[a]For 1987-88.

listed in Table 4–1, separately for blacks and whites. When the width of the sex gap is used to rank the causes, the list is quite different from that in Table 4–1. For example, males are nearly four times more likely than females to commit suicide, and that cause has the highest sex mortality ratio; males are also three times more likely to be homicide victims. In fact, not until we reach diseases of the heart, in sixth place on this list, does the death rate of males drop slightly below twice that of females. The rates are fairly close together only near the bottom of the list—for cerebrovascular diseases and diabetes. Thus, environmental causes still take a far heavier toll of males than females, though biological differences continue to play a significant part in other causes. Males die at inordinately high rates because of suicide, aggressiveness and other violence that results in murder, recklessness and job risks that cause accidents, and alcohol abuse that leads to liver diseases. Pulmonary diseases, such as emphysema and asthma, also take a much heavier toll of males, partly because men are still more likely than women to be long-term smokers, although the sex difference in smoking behavior has decreased to the advantage of men and disadvantage of women. The death rate for heart disease among men is twice that among women, so while changes in eating habits, smoking, exercise, medical procedures, and other manipulable conditions have greatly reduced the death rate for heart disease, they have not altered the sex gap as much. Just between 1980 and 1988, for example, while the age-adjusted death rate for heart disease fell 20 percent among males and 15 percent among females, the sex gap narrowed only 5 percent.

The situation for cancer reflects other dynamics. The death rate among males for cancer dropped slightly between 1980 and 1988, while the rate among females rose slightly, so the sex gap narrowed by 3 percent. This reflects a discouraging lack of progress in preventing and treating some kinds of malignancies and increased susceptibility of women to lung cancer, even while the death rate for other forms has fallen. For instance, the age-adjusted death rate among males for cancer of the respiratory system remained almost unchanged in the 1980s, but it rose 33 percent for females. Women's death rate for breast cancer fluctuated somewhat, but it was slightly higher in 1988 than in 1980 and not much different than in 1950. In some age groups, however, it fell significantly, while in others it rose sharply, as discussed in Chapter 5. The death rate for prostate cancer rose somewhat in the 1980s and significantly over earlier decades. The rate for cancer of the lip, oral cavity, and pharynx fell among both sexes, but more among males than females, as did cancer of the urinary system. On the other hand, the death rate for malignancies of the digestive system dropped more among females than males, while the rate for genital cancer rose slightly among males and dropped 11 percent among females, partly because of greater use of the early detection "Pap smear." The net result of these changes in the 1980s was a 2 percent overall decline in the age-adjusted death rate for cancer among males and a 2 percent rise among females, producing a modest decrease in the male/female mortality ratio for that cause.

The sex mortality ratio is larger among blacks than whites for all causes but three: heart disease, including ischemic heart disease; chronic liver disease and cirrhosis; and diabetes. Since heart disease kills such a large number of members of both races and the death rate for diabetes is much higher among females than males, the comparatively low sex ratios for those two causes among blacks help counterbalance their high ones for many other causes and produce an overall ratio close to that for whites. Blacks have high sex mortality ratios for suicide, homicide, accidents and several other causes in Table 4–2. Many reflect the high risks that accompany being a black male in America, especially if one lives in an inner city, is an adolescent or young adult, falls below the poverty line, is unemployed, has a substance-abuse problem, is vulnerable to violence, and endures other disadvantages. Such males are truly marginalized, and though the plight of poor black females is serious, black males have higher death rates for most major causes than do black females and whites of both sexes. Therefore, concern for poor women and children, regardless of race, should not obscure the excessively high death toll among black men and their need for amelioration.

Some efforts to lower mortality among black males, especially in the inner city, might include improved rehabilitation of drug and alcohol abusers and treatment for those who are mentally ill, along with better educational and occupational opportunities to help black males afford health care and minimize risk-taking behavior. Most needed, however, are comprehensive social changes and behavior modification techniques that provide viable alternatives to violence, drug and alcohol abuse, and other dangerous behaviors. In turn, such modification requires more positive black role models for inner-city boys and young men, ways to reduce anomie and hopelessness so as to counteract the compelling macho norms of delinquent gangs, dispersal of young black males away from the concentration of deviance in many inner-city neighborhoods, revival of viable family structure and its constraints on narcissistic and antisocial behavior, and a host of other measures. All are difficult and require fundamental socioeconomic and psychosocial changes that finally obliterate the effects of racism, its centuries-long marginalization of many black males, and its adverse effects on personal health habits and behavior.

Table 4–2 also shows sex mortality ratios for HIV, which produces AIDS among many who are infected. As noted earlier, the NCHS began using a separate reporting category for HIV infections in 1987, so the figures in the table are only for 1987–88. Nevertheless, they show that white males are 13 times as likely as white females to die of HIV infections and that black males are 5.2 times as likely as black females to succumb. The age-adjusted death rate for this cause, however, is highest by far among black men: in 1987–88 it was 28.5, making it their sixth leading cause of death. By comparison, the death rate was 9.1 among white males, 5.5 among black females, and 0.7 among white females.[29] The relative disadvantage of black men is clear in this case too, but data for all groups show how rapidly the HIV/AIDS epidemic is spreading. In fact, the two-year averages for all groups consist of much higher age-adjusted death rates in 1988 than in 1987. Moreover, the disease is spreading rapidly among teenagers, though their number of cases is still relatively small. AIDS continues to progress most rapidly, however, among men in their 20s and 30s.

DIFFERENTIALS BY RACE AND HISPANIC ORIGIN

Blacks and Whites

To ascertain how well blacks of each sex fare relative to whites from mortality for each major cause, Table 4–3 shows the black/white age-adjusted mortality ratio by cause. These data substantiate earlier observations and show the striking relative disadvantage of blacks in deaths for eight of the ten major causes. In addition, the ranking of causes by the black/white mortality ratio differs from

Table 4–3
Black/White Ratio of Age-Adjusted Death Rates for Selected Causes, by Sex, 1986–88

Cause	Both Sexes	Male	Female
All causes	152.6	153.4	152.9
Homicide and legal intervention	607.4	708.9	424.1
Diabetes mellitus	235.6	198.9	263.4
Cerebrovascular diseases	184.3	189.8	178.7
Chronic liver disease and cirrhosis	175.7	175.2	176.9
Pneumonia and influenza	145.4	156.3	127.7
Diseases of heart	138.1	127.3	156.5
Ischemic heart disease	103.4	92.6	122.7
Malignant neoplasms	132.2	144.0	119.8
Respiratory organs	127.4	144.0	100.8
Digestive organs	151.4	157.4	146.3
Accidents and adverse effects	124.9	134.7	115.0
Motor vehicle	91.5	102.1	77.9
Chronic obstructive pulmonary diseases	81.4	89.6	68.8
Suicide	53.6	58.7	43.4
Human immunodeficiency virus infections[a]	328.6	313.2	785.7
All other causes	196.7	196.5	196.8

Source: National Center for Health Statistics, Vital Statistics of the United States, 1986, 1987, Vol. II, Mortality, Part A (Washington, DC: Government Printing Office, 1988-90), various tables; NCHS, Health, United States, 1989 (Washington, DC: Government Printing Office, 1990), Table 23; NCHS, "Advance Report of Final Mortality, Statistics, 1988," Monthly Vital Statistics Report 39:7 (Washington, DC: Government Printing Office, November 1990), Tables 12, 25.

[a]For 1987-88.

the ranking according to the age-adjusted death rate or the male/female mortality ratio.

Homicide ranks at the top of the list, with blacks six times more likely than whites to be murdered, although males fare significantly worse than females. Blacks are twice as likely as whites to die of diabetes; over 1 million blacks are afflicted and both the number and incidence of diabetes cases have increased substantially.[30] Blacks are nearly twice as likely as whites to succumb to strokes due to brain hemorrhage and thrombosis and other occlusion. One reason is that much higher percentages of blacks than whites suffer hypertension, and while it is not among the ten leading causes for either race, the death rate for it as an underlying cause is 3.8 times greater among blacks than whites. Blacks are much more likely to die of liver disease and cirrhosis and of pneumonia and influenza. Heart disease is sixth in the ranking by the black/white mortality ratio, though the racial gap is greater for women than men. Moreover, while blacks are only somewhat more likely than whites to be struck down by ischemic heart disease, they are 4.7 times more likely to die of hypertensive heart disease, again implicating high blood pressure. In fact, if hypertension could be well controlled among blacks, the effect in lowering their death rate would rank with comparable reductions in violence, smoking, and substance abuse.

The black/white ratio shows that blacks are one-third more likely than whites to die from cancer, though the gap is considerably greater for males than females. Black males are at much greater risk than white males for cancer of the respiratory system, mostly lung cancer, but among females there is little racial difference. For malignancies of the digestive system and peritoneum, which includes colorectal, stomach, liver, gall bladder, and pancreatic cancer, black males and females fare considerably worse than their white counterparts. Black women have a 16 percent higher death rate than whites for breast cancer; they also have higher rates for cancer of the uterus and cervix, but a lower one for ovarian cancer. Black men are considerably more likely than white men to die of prostate cancer. A black person of either sex is less likely than a white to fall victim to brain cancer or lymphoma or to fast-moving melanoma, which is retarded because dark skin pigmentation lowers susceptibility to sun damage.

Accidents claim enough blacks to keep the ratio above 100, but motor vehicles are implicated in a much larger share of deaths of white than black women. Other accidents are far riskier to blacks, especially those involving drowning, fire, and drugs.

Blacks are considerably less likely than whites to succumb to chronic obstructive pulmonary diseases, such as emphysema and asthma, and they are especially unlikely to commit suicide, producing a rate only half that among whites.

Hispanics and Non-Hispanic Whites

As suggested in Chapter 2, it is difficult to compute age-adjusted death rates for the Hispanic population because deaths by age are reported for only about half the states and the District of Columbia, and the Census Bureau does not provide intercensal estimates of the Hispanic population by ten-year age groups for those same states, though it does for the nation as a whole. Thus, the two data sets necessary to calculate age-adjusted death rates, whether by the direct or indirect methods of standardization, are not comparable. Nevertheless, it is possible to rank causes of death according to the percentage of deaths for which each is responsible. Table 4–4 shows the 20 leading causes of death for the two groups in 1987 and the percentage of all deaths produced by each cause.

Heart disease is the principal cause of death among both Hispanics and non-Hispanics, but it accounts for a much smaller share of all deaths among Hispanics. Malignant neoplasms are second, but in this case too the percentage of all deaths caused by cancer is smaller among Hispanics than non-Hispanics, partly because of age profiles. Cancer deaths are fairly heavily concentrated in the older ages

Table 4–4
Percent of All Deaths Attributable to Leading Causes, by Hispanic and White Non-Hispanic Origin, 1987

Cause	Hispanic Rank	Hispanic Percent	White Non-Hispanic Rank	White Non-Hispanic Percent
All causes	...	100.0	...	100.0
Heart disease	1	25.4	1	37.2
Malignant neoplasms	2	16.9	2	22.8
Digestive organs	2A	4.9	2B	5.7
Respiratory organs	2B	3.1	2A	6.5
Genital organs	2C	2.0	2C	2.4
Breast (female)	2D	1.5	2D	2.0
Leukemia	2E	0.8	2F	0.8
Urinary organs	2F	0.6	2E	1.0
Accidents and adverse effects	3	9.3	5	3.9
Motor vehicle	3A	5.4	5A	2.0
Other	3B	3.9	5B	1.9
Cerebrovascular diseases	4	5.5	3	7.0
Homicide	5	4.6	17	0.5
Chronic liver disease and cirrhosis	6	3.4	11	1.2
Symptoms, signs, ill-defined conditions	7	3.1	9	1.4
Pneumonia and influenza	8	3.0	6	3.5
Diabetes mellitus	9	3.0	7	1.6
Human immunodeficiency virus infections	10	2.9	14	0.6
Certain conditions of perinatal period	11	2.7	15	0.5
Chronic obstructive pulmonary diseases	12	2.1	4	4.1
Suicide	13	1.9	8	1.5
Congenital anomalies	14	1.9	16	0.5
Nephritis, nephrotic syndrome, and nephrosis	15	1.2	12	1.0
Septicemia	16	1.1	13	0.8
Atherosclerosis	17	0.6	10	1.2
Benign neoplasms	18	0.3	19	0.3
Ulcer of stomach and duodenum	19	0.3	18	0.3
Meningitis	20	0.1	20	a
All other causes	...	10.7	...	10.1

Source: National Center for Health Statistics, Vital Statistics of the United States, 1987, Vol. II, Mortality, Part A (Washington, DC: Government Printing Office, 1990), Sec. 1, Tables 35, 41.

aLess than 0.1 percent.

and the comparatively large Mexican and Puerto Rican segments are relatively "young" groups, although the Cuban contingent is more an aging population. Accidents are third among Hispanics, but only fifth among non-Hispanics, and while vehicular and nonvehicular accidents account for about equal proportions of deaths among non-Hispanics, motor vehicle deaths clearly dominate the accident category among Hispanics, owing partly to the relatively high proportion of teenagers and young adults. Cerebrovascular diseases are fourth among Hispanics and third among non-Hispanics. Homicide is fifth for Hispanics but a distant seventeenth for white non-Hispanics, and the percentage of deaths it causes among Hispanics is nine times the percentage among non-Hispanics. A similar though smaller gap exists for chronic liver disease and cirrhosis, partly because of differences in alcohol consumption.

The category "symptoms, signs, and ill-defined conditions" is seventh on the list among Hispanics and ninth among non-Hispanics. It is not a cause at all, of course, but a catchall category that is more important for Hispanics because more of their causes of death are poorly identified and reported. Pneumonia and influenza, in eighth place, account for a smaller share of deaths in the Hispanic than the non-Hispanic population, again partly because of variable age profiles. Diabetes, ranked ninth for Hispanics and seventh for non-Hispanics, causes about twice the proportion of deaths among Hispanics.

HIV infections are tenth as a cause among Hispanics and only fourteenth among white non-Hispanics. Thus, in 1988 Hispanics were 7 percent of the nation's population aged 13 and over but had 13 percent of the reported AIDS cases in those ages and 13 percent of the deaths for that cause; HIV/AIDS is increasing rapidly as a problem among Hispanics. Blacks, however, were 11 percent of the population aged 13 and over and had 29 percent of the AIDS cases and 29 percent of the AIDS deaths, so the problem is even greater for them than for Hispanics. Regardless of race or ethnic group, HIV/AIDS is far more common among males than females.

Hispanics are far more likely than white non-Hispanics to die from certain conditions originating in the perinatal period and from congenital anomalies, even though their reported infant mortality rate is not much different than among white non-Hispanics. The proportional losses of Hispanic children after infancy, however, are considerably greater than among their non-Hispanic counterparts from causes such as leukemia, congenital heart problems, pneumonia and influenza, septicemia, accidents, and homicide.

Some causes, in addition to heart disease, cancer, and stroke, are considerably less important proportionately for Hispanics than non-Hispanics, including chronic obstructive pulmonary diseases, which rank only twelfth as a cause of death for Hispanics but fourth for white non-Hispanics, partly because of differences in the age profile, but also because of differences in smoking. Suicide also is much less common among Hispanics, partly because the Roman Catholicism that many espouse strongly condemns it, and partly because kinship and friendship groups protect against anomie and other psychosocial states that can lead to suicide. Death from atherosclerosis also is less likely among Hispanics, partly because of age differences, but also because many recent Hispanic migrants to the United States have not yet adopted the high-fat diet that feeds the disease.

Within the several specific ten-year age groups for which deaths are reported, the ranking of many causes is about the same for Hispanics and non-Hispanics. Some differences persist, however: HIV infections rank much higher among Hispanics than non-Hispanics at all ages under 65.[31] Homicide also ranks higher for Hispanics in all age groups from 15 to 65, and chronic liver disease and cirrhosis are proportionately more lethal for Hispanics in all ages 25 and over.[32]

A comprehensive study of Hispanic death rates for 1979–81, edited by Ira Rosenwaike, shows not only that Hispanic males as a whole have lower death rates than non-Hispanic males for heart disease, cancer, and stroke, but so do the Mexican, Puerto Rican, and Cuban contingents of the overall Hispanic group.[33] On the other hand, all three groups of Hispanic males have substantially higher death rates for homicide than the total male population of the United States. For all other causes, however, the death rates among Hispanic males of Mexican, Puerto Rican, and Cuban backgrounds compare favorably with those of non-Hispanic males.[34] The death rates among Hispanic females show that they are somewhat less likely than non-Hispanic females to die of cancer and suicide, but for many other causes they are at a relative disadvantage.[35] Therefore, the mortality rate for both sexes combined reflects the proportional advantages and disadvantages discussed above.

The fairly advantageous mortality situation for many causes that does appear among Hispanics seems partly due to the fact that a sizable share of that population consists of first-generation immigrants, who tend to be healthier than a cross-section of people in their home countries because of the selectivity of immigration. It also seems partly due to differences in diet and other lifestyle factors, and to the fact that some migrants, especially Cubans, came disproportionately from the middle and upper classes.[36] Mortality patterns among elderly Hispanics suggest that Spanish-background migrants, not unlike Japanese and Chinese migrants to the United States, "have retained the more favorable aspects of their original environment, thereby holding their mortality to a level lower than that generally prevailing in the United States."[37] Logically, therefore, as second-generation and longer-term Hispanics assimilate more fully into American culture, their mortality patterns come to resemble more closely those of white non-Hispanics, so both the mortality advantages and disadvantages of the Hispanic group tend to diminish with each generation.[38] This convergence too is reflected in the proportional significance of many of the causes of death among Hispanics discussed earlier.

DIFFERENTIALS BY AGE

The age-adjusted death rate enables comparability but does not show the relative importance of various causes in given age groups. The age-adjusted rate cannot indicate, for example, that heart disease accounts for less than 4 percent of deaths among children aged 1–4 but 46 percent of deaths among persons 85 and over. Thus, the analysis turns to age-specific death rates for 15 leading causes, ranked according to age-adjusted death rates and using seven broad age categories. Several ten-year age categories are grouped because their patterns of death by cause are similar, though the rates for virtually all causes rise with age. Some observations about infants not made in Chapter 3 also are included here. Table 4–5 provides the necessary data for the following sections.

Infants Under 1 Year

The infant death rate is quite high compared with that of children older than 1 year, and various conditions of the perinatal period account for almost half the losses. (The infant death rate is the number of deaths under 1 year per 100,000 in that age group, whereas the infant mortality rate, used in Chapter 3, is the number of infant deaths under 1 year per 100,000 or 1,000 live births.) Heart disease and accidents are responsible for another sizable share, and so are pneumonia and influenza, followed by homicide, septicemia, and nephritis and related conditions. Septicemia is "blood poisoning" that results from bacterial infection and causes death from the bacteria themselves or from toxins they produce. Infants are less able than older children to resist septicemia, especially if they are premature and underweight and their immune systems are weak or they do not receive antibiotics. Nephritis, nephrotic syndrome, and nephrosis make up a category of kidney infections and disorders in which most deaths come from renal failure or impaired function, often because of small kidneys. Low birth weight and immaturity are often implicated in these causes.

The death rates of infants are high relative to those of older children for each cause except cancer and diabetes. In fact, the infant death rates for heart disease, cerebrovascular diseases, and chronic obstructive pulmonary diseases are higher than those for all age groups until 35–44; the infant rates for influenza and pneumonia and nephritis and related diseases are not reached again until ages 55–64; HIV infections take a higher toll of infants than other age groups through 15–24. To the extent these are preventable causes, the risky lifestyles of many mothers and fathers, inadequate prenatal care, poor safety precautions, and other factors discussed in Chapter 3 keep the overall infant death rate relatively high; the rates for some causes are even close to those among middle-aged adults.

Children Aged 1–4 Years

Relatively few deaths from any cause occur in the ages 1–4, although the overall death rate is still about twice that among children aged 5–14. Accidents are the leading cause by far, although nonvehicular accidents, such as suffocation, burning, and drowning, kill about twice as many children as those involving vehicles. Accidents and several other causes reflect the child's environment and

Table 4–5
Age-Specific Death Rate for 15 Leading Causes, 1986–88

Cause[a]	0–1 Year	1–4 Years	5–14 Years	15–24 Years	25–34 Years	35–44 Years	45–54 Years	55–64 Years	65–74 Years	75–84 Years	85+ Years
All causes	1,019.6	51.5	25.8	101.3	133.6	215.5	496.3	1,244.0	2,760.8	6,317.3	15,437.9
Diseases of heart	24.6	2.4	0.9	2.8	8.4	35.8	138.8	411.3	1,011.7	2,580.1	7,117.0
Malignant neoplasms	2.5	3.8	3.3	5.2	12.5	44.3	163.5	446.2	844.4	1,299.7	1,623.0
Accidents	24.5	20.1	12.3	49.9	38.7	31.8	30.6	35.1	49.4	106.5	257.1
Cerebrovascular diseases	3.4	0.4	0.2	0.7	2.2	7.0	19.9	52.2	158.7	563.3	1,734.4
Chronic obstructive pulmonary diseases	1.5	0.3	0.3	0.5	0.6	1.7	9.4	47.8	148.9	302.7	378.0
Pneumonia, influenza	17.3	1.4	0.4	0.7	1.8	3.6	7.1	18.4	58.4	244.7	1,053.0
Suicide	c	c	0.7	13.1	15.5	15.0	15.6	16.4	19.2	25.6	21.0
Dibetes mellitus	0.1	0.1	0.1	0.3	1.5	3.7	9.9	27.0	60.8	123.4	216.5
Chronic liver disease and cirrhosis	0.5	c	c	0.2	2.6	9.9	20.0	32.1	35.8	32.2	20.6
Homicide	7.6	2.5	1.2	14.5	15.7	11.0	7.7	5.4	4.3	4.6	4.8
Certain conditions of perinatal period	476.8	0.9	c	c	c	c	c	c	c	c	c
HIV infections[b]	2.2	0.8	0.2	1.4	12.7	15.8	5.9	3.8	1.5	0.8	0.5
Nephritis, nephrotic syndrome, nephrosis	6.0	0.1	0.1	0.2	0.6	1.3	3.3	9.4	26.3	78.9	217.2
Septicemia	7.0	0.6	0.1	0.3	0.7	1.4	3.2	9.0	23.6	67.3	189.8
Atherosclerosis	c	c	c	c	c	0.1	0.7	3.9	15.8	72.8	411.7

Source: National Center for Health Statistics, "Advance Report of Final Mortality Statistics," 1986, 1987, 1988, Monthly Vital Statistics Report (Washington, DC: Government Printing Office, 1988–90), Table 5.

aDeaths per 100,000 population in specified group.

bFor 1987–88.

cLess than 0.1 percent.

the situation of its parent(s) or others who provide care. Most illnesses suffered by young children, however, are not chronic but "episodic and short term," so poor health among children is better reflected in poor dental care, uncorrected vision and hearing impairments, nonfatal accidental injuries, abuse, poor nutrition, and other deficiencies apart from the causes of death.[39] The extent to which children are immunized against various illnesses also indicates their well-being and is related to the educational levels of their parents, especially mothers, since education is the factor most associated with understanding and seeking immunization.[40] Furthermore, elements of health, like causes of death, are so interrelated that when one disease is eliminated or diminished, child morbidity from several other diseases is also likely to improve and the death rate to fall.[41]

Only 5 percent of the deaths in the 1–4 age group is from infections and parasitic diseases—causes that took a much heavier toll of young children early in the century. Widespread immunization has greatly reduced deaths from "measles, mumps, rubella [German measles], polio, diphtheria, pertussis [whooping cough], and tetanus," although cases of measles, mumps, and pertussis still number several times more than the Surgeon General's objectives.[42] They and others can also stage a comeback if many parents fail to immunize their children. Among children who do die from infectious and parasitic diseases, 26 percent succumb to HIV infections, 22 percent to septicemia, and 19 percent to meningococcal infections. Thus, even though these preventable causes take a small number of children's lives, they claim two-thirds of the ones who do die. Moreover, the incidence of HIV/AIDS is rising among children under 5: the number of cases in 1989 was more than twice the number in 1988; black children accounted for 57 percent, Hispanics for 25 percent, and whites for 18 percent.[43]

The lingering effects of congenital anomalies, not shown in the table, are the second leading cause of death for children 1–4. Cancer is third, and apart from the many cases in which the site is unspecified, leukemia is the chief culprit, causing about a third of the cancer deaths. Young children also are victims of homicide (fourth place) and heart disease (fifth place), which often results from congenital anomalies. The other causes taper off to very low numbers and rates. In fact, while there were 14.6 million children aged 1–4 in 1988, there were only 7,429 deaths, and though that number can be reduced further, it represents substantial progress in dealing with most causes that victimize young children. Congressional action in 1991 to cover all poor children aged 0–6 under Medicaid health insurance by 2003 should produce even more progress for infants and children.

Children Aged 5–14

Youngsters 5–14 have the lowest overall death rate of any age group and some causes of death differ markedly from those of younger children. Accidents account for almost half the losses in the ages 5–14, and those involving motor vehicles take a larger toll than all other accidents combined. Cancer moves to second place, and apart from malignancies at unspecified sites, leukemia is still the major type. Congenital anomalies account for enough deaths to place them third; homicide continues in fourth place and heart disease in fifth. Suicide becomes the sixth leading cause, starting an increase in the rate that rises until middle age, tapers off briefly, and then increases steadily into the oldest ages. The number of childhood suicides is small, but it implies much about the problem conditions under which many children live, the abuse of drugs and alcohol by some older members of this age group, and even the suggestibility of young people to suicide as a way of coping with reality. Indeed, there is a "psychosocial contagion" about suicide, though it is greater among adolescents than children, and news of one suicide case, especially if it involves multiple victims, may cause other cases to "cluster" in time and place.[44]

The ages 5–14 are the absolute low point in the death rates for nearly all 15 leading causes, which subsequently take a rising toll. In fact, the death rates for several causes among persons aged 15–24 jump to several times those among children aged 5–14. Cancer has a lower death rate among infants under one than among children 5–14, and, of course, suicide first appears for the older group. But even the death rate for HIV/AIDS is lower among them than among younger children because those 5–14 probably would not have survived to that age had they been infected at birth, whereas they are generally too young to contract the diseases through sexual contact or intravenous drug use.

Adolescents and Young Adults Aged 15–24 Years

The death rate among adolescents and young adults jumps dramatically and certain causes begin to take a high toll, though except for accidents the rate for each cause is still lower than among all age groups past 24. In fact, accidents stand out sharply as the major cause, rising to four times the rate among children 5–14 and accounting for almost half the deaths of persons 15–24. Motor vehicle accidents are responsible for 76 percent of the accidental deaths and 38 percent of all deaths. The members of this age group now have driver's licenses, and their number of deaths and death rate for motor vehicle accidents are substantially higher than in any other ten-year age group. In fact, while persons aged 15–24 were 15 percent of the population in 1988, they suffered 30 percent of the deaths by motor vehicle. The incidence is greatest among white males, and though they were only 6 percent of the population, they had 19 percent of the motor vehicle deaths. Alcohol is involved in about half the cases and drugs also play a role, for these are the years when people are especially likely to experiment, take risks, and otherwise grope toward adulthood.

Homicide and legal intervention is the second leading cause of death in this age group, though the number of deaths and the rate are greater in the ages 25–34. Homicide is the leading cause of death among black males aged 1 whose rate is far higher than that of any other group: 85.6, compared wi among black females, 11.2 among white males, and 3.9 among white

In the ages 20–24 the homicide rate among black males (112.6) is nearly twice that in the ages 15–19.

Suicide is the third leading cause of death, with a rate fairly close to that for homicide. White males are more likely than any other race or sex group to commit suicide, and within the overall age group, those 20–24 have a higher rate (27.5) than the ones 15–19 (17.6). Thus, violent deaths, whether by murder, motor vehicles, or self-destruction, account for 77 percent of all deaths in these ages. Apart from these three, therefore, the death rates for other causes are relatively low, although the loss to cancer begins its inexorable rise toward the older ages. So do deaths for HIV/AIDS, which ranks sixth among the causes. Its death toll is highest among black males, with a death rate in 1988 of 31.6 for HIV/AIDS, compared with 9.9 among white males, 6.2 among black females, and 0.7 among white females.

The Surgeon General's 1990 goal for the 15–24 age group was no more than 93.0 deaths per 100,000, but despite progress toward that goal, especially in 1980–83, by the end of the 1980s it went unmet. The goal of a suicide rate not exceeding 10.9 was not met either, nor was a homicide rate of 60.0 or less among black males. In fact, while their homicide rate fell somewhat between 1980 and 1984, it rose sharply in each succeeding year. The overall mortality goal went unmet partly because persons aged 11–20 are less likely than children under 11 or adults over 20 to get health care, even though they have a higher incidence of acute conditions than older age groups. Those conditions are largely self-limiting and many do not require a doctor's care, but some that do are not properly treated, and the rising death rate from ages 5–14 to 15–24 reflects that neglect.[45]

As suggested in connection with motor vehicle fatalities, drugs and alcohol are implicated in many violent deaths in the ages 15–24. Nevertheless, there is more awareness, especially among high school seniors, of the dangers in abusing marijuana, cocaine, and alcohol, so use of those substances appears to have declined somewhat.[46] The Surgeon General's objective of 80 percent of seniors perceiving great risk in using various substances has about been realized for marijuana and the regular use of cocaine, including crack cocaine, though there is still some distance to cover for barbiturates and especially cigarettes and alcohol. In fact, alcohol abuse seems to have risen on college campuses as the use of other drugs has fallen. Moreover, most improvements occurred in the middle and upper classes. The substance-abuse problem remains as serious as ever among poor young people in the inner city, a large share of them black; among the homeless, especially mentally ill persons released from institutions without follow-up treatment; and among other "social casualties" strongly addicted to drugs and alcohol.

Adults Aged 25–34 Years

In the ages 25–34 the overall death rate is a third again as high as in the ages 15–24. Accidents still account for over twice as many deaths as any other cause, and homicide remains in second place, suicide in third, though the rates for the last two are almost the same. The rates for accidents and homicide are higher in the ages 24–29 than 30–34, but there is little difference for suicide. Moreover, the patterns by race and sex for the three leading causes are similar to those in the ages 15–24.

HIV infections move up to fourth place. That rank is higher than in any younger ages; it persists in the ages 35–44 and then drops off significantly. Though persons 25–34 make up about 18 percent of the population, they account for 37 percent of the HIV deaths. The losses are especially heavy among black males, and while they are only 6 percent of the total population aged 25–34, they account for 26 percent of the HIV deaths in those ages. White males make up 42 percent of the 25–34 age group and incur 61 percent of the HIV deaths. Black females are 7 percent of the population aged 25–34 and experience 7 percent of the HIV deaths; white females are 41 percent but have only 5 percent of the deaths from that cause.

Degenerative illnesses now move up in the list. Cancer is in fifth position and lymphatic cancer kills more members of this age group than any other malignancy. Breast cancer, which is a complex disease that usually develops gradually, is the chief cause of cancer deaths for both black and white women, while white men are most likely to succumb to lymphatic cancer and black men to cancer of the digestive system. Respiratory cancer ranks only fifth for men of both races and sixth for women, largely because cigarette smoking has not yet had time to take its heaviest toll, although lung cancer does become increasingly important with age. Death rates for cancer in the ages 25–34 are nearly the same among black and white women and men. Women gain the advantage in the mid-40s, however, and the sex gap continues to widen with age.

Heart disease is sixth and the principal form is ischemic heart disease, mostly from acute or old myocardial infarction. Unlike cancer, heart disease takes a higher toll of males than females of both races in this and every other age group, from infancy to advanced old age. Among whites aged 25–34 the rate is already 2.3 times higher for men than women, and among blacks it is 1.8 times greater.

Chronic liver disease and cirrhosis, followed by cerebrovascular diseases, pneumonia and influenza, and diabetes, round out the list of the ten major killers in this age group. Chronic obstructive pulmonary diseases, septicemia, and nephritis and related diseases all still have death rates below 1.0 per 100,000, but they have risen from the younger age group and continue to do so into the older ages.

Adults Aged 35–64 Years

Patterns of death by cause are fairly well established by age 35, so we can look at a 30-year category. The death rate for all causes more than doubles in each succeeding ten-year age group, but among persons aged 35–64 cancer becomes the major cause, and its death rate multiplies by ten from one end to the other of the age group) Respiratory and intrathoracic malignancies are the chief types of cancer, causing about 23 percent of cancer deaths among persons aged 35–44 and 35 percent among those 55–64. Moreover, cancer of the lung, bronchus, and trachea cause about 96 percent of all deaths from respiratory cancer. In fact, lung cancer is the most common malignant neoplasm in America's population as a whole, and among black and white women its incidence is increasing faster than that of any other type of cancer.[47]

Despite the major role of lung cancer, the specific malignancies that kill men and women aged 35–64 vary substantially. Among men of both major races respiratory cancer does rank first, but among white and black women aged 35–49 breast cancer heads the list. In the ages 50–64 it drops to second place for white women and respiratory cancer ranks first, while for black women breast cancer falls to third place and cancer of the digestive system and of the respiratory system rank first and second, respectively.

The deaths of women from breast cancer are sufficient to push the overall cancer death rate among white women aged 30–49 above the rate among white men. The same is true in the black population in all age groups from 25 to 39. In addition, the death rate for genital cancer is higher among white women than white men in all age groups from 25 to 64, and it is higher among black females than males in all age groups from 10 to 39. For all other forms of cancer the death rate is considerably lower among women than men, race aside, so breast cancer and genital cancer have an especially powerful impact on the death rates of women from their late 30s to their early 60s. The death rate for breast cancer continues to rise with age and is highest among women aged 85 and over, but it takes its largest proportional toll earlier.

Within the category of genital neoplasms, prostate cancer accounts for 90 percent of the total among white men aged 35–64 and 97 percent of the total among black men. In the ages 35–49 it is responsible for only 39 percent of the deaths from genital malignancies among white men, but in the ages 50–64 it causes 95 percent of the deaths. The comparable figures for blacks are 71 percent and 98 percent, respectively. Among white women aged 35–64 the leading genital malignancy is ovarian cancer, which accounts for 44 percent, followed by cervical cancer; among black women the order is reversed and cervical cancer causes 45 percent of the deaths.

In the category of digestive cancer, colorectal malignancies are the principal cause among males and females of both races and in all age groups that make up the 35–64 category. Colorectal cancer causes 34 percent of the losses among white men aged 35–64, followed by cancer of the pancreas, the stomach, the esophagus, and the liver and intrahepatic bile ducts. Colorectal cancer causes 31 percent of the deaths from digestive cancer among black men, 52 percent among white women, and 43 percent among black women, followed in variable order by the other digestive malignancies noted for white men. The relative virulence of pancreatic cancer among black women is tied to their comparatively high incidence of diabetes mellitus, the noninsulin-dependent form of which occurs most often among overweight middle-aged and elderly women. Since black women aged 35–64 are about twice as likely as white women in those ages to be overweight, they are also more apt to suffer from diabetes and other disorders of the pancreas, including cancer. This is also one reason the death rate among black men for diabetes is twice that among white men, since black men are significantly more likely to be obese, though less frequently than black women.[48]

The second leading cause of death in the ages 35–64 is heart disease, though it becomes the major killer for males and females at various ages. In 1988, for example, heart disease was the leading cause of death for both black and white men in the ages 40–44, surpassing accidents. Heart disease did not lead the causes among white women until the ages 70–74, when its death rate finally passed the one for cancer. Among black women heart disease became the leading cause in the ages 60–64, when it outdistanced cancer. There is also considerable change between the ages 35–49 and 50–64 in the relative importance of heart disease as a cause of death among blacks and whites. In the first age group the death rate for heart disease among black males is 2.1 times that among white males, whereas in the older group it is only 1.5 times greater. Among females the rate decreases from 3.3 times to 1.6 times higher. Ischemic heart disease remains the major form of heart disease in all these ages and for both races and sexes, involving either acute or subacute myocardial infarction or old infarction. Among blacks, however, hypertensive heart disease is proportionately more important than among whites.

Accidents are the third leading cause of death until about age 55, after which they drop to fifth, only to be succeeded in third place by stroke and other cerebrovascular diseases, the death rates for which rise rapidly. Among the causes in Table 4–4, HIV infections remain in fourth place for persons aged 35–44, when HIV reaches its peak death rate. It then drops to eleventh position in the ages 45–54 and the death rate for it tapers off with age. Suicide also falls in the ranking and the death rate for that cause is nearly level across the 35–64 age group. The death rates for chronic obstructive pulmonary diseases, pneumonia and influenza, diabetes, nephritis and related diseases, septicemia, and atherosclerosis all rise dramatically from age 35 to 64, while the death rate for homicide is cut nearly in half.

The racial mortality crossover first appears in the ages 35–64 for certain causes. Discussed in Chapter 2, this phenomenon results from a reported lower death rate among blacks than whites at the older ages, thus reversing the usual black/ white mortality ratio. For all causes combined, it generally does not occur until age 80 or over; in 1988, for example, it took place in the ages 85 and over for both males and females. The crossover in the oldest ages, however, results from what first happens to certain causes of death in the 70s, 60s, and even 50s.

Table 4–6 shows the categories and subcategories of causes for which the crossover occurs and the ages when it first appears. Causes not on the list are those for which the death rate of blacks remains above that of whites at every age (e.g., homicide), the ones for which the death rate of blacks is lower than that of whites at early ages but then rises higher (e.g., genital cancer among males), and the one major cause (suicide) for which the rate is significantly lower among blacks than whites at every age. Thus, the crossover takes place among men somewhere between ages 50 and 69 for ischemic heart disease, cancer of the urinary organs, leukemia, and chronic obstructive pulmonary diseases. The crossover among women occurs somewhere between ages 55 and 59 for neoplasms of the respiratory organs, the breast, and the urinary organs, and for leukemia, motor vehicle accidents, and chronic obstructive pulmonary diseases. For all other causes in Table 4–6 the crossover occurs later, but the list is fairly long and we can infer that while some of the crossover is a statistical fiction because of faulty data for persons 85 and over, it also seems to be a real phenomenon that starts at various younger ages depending on the cause of death.

Adults Aged 65 and Over

At 65 and over the ranking of causes changes in various ways. Diseases of the heart move into first place and remain there and cancer shifts to second position, except among persons 85 and over, for whom cerebrovascular diseases take second place. Pneumonia and influenza also move up in the list with increasing age, as do chronic liver disease and cirrhosis. Atherosclerosis gains rapidly as a cause of death, moving from twelfth place among persons 65–74 to fifth among those 85 and over. In fact, the death rate for atherosclerosis is 26 times higher in the older than the younger of those age categories. Eventually, it becomes a major underlying cause of death among persons whose arteries are heavily blocked, while it still contributes to death from heart disease and stroke. Chronic obstructive pulmonary diseases, diabetes mellitus, and chronic liver disease and cirrhosis become relatively less important in the older ages, although the death rates for the first two rise dramatically. The death rate itself falls for liver disease and cirrhosis, HIV infections, and, after age 84, for suicide. The death rate for homicide rises modestly from 65 to 85 and over because the very old are especially vulnerable to violence, whether at the hands of spouses, adult children, nonfamily caregivers, or street hoodlums.

The proportion of all deaths from the three leading causes—heart disease, cancer, and stroke—change considerably in the elderly population. Deaths from heart disease account for a steadily increasing percentage of all losses in each succeeding age group over 65 regardless of race and sex, even though the death rate for that cause fell 11 percent between 1978 and 1987 (see Figure 4–1). Thus, among persons aged 65–69 heart disease causes 36 percent of the deaths, but among centenarians it causes 52 percent. Conversely, cancer causes a decreasing percentage of all deaths among blacks and whites of both sexes. In the group aged 65–69 it causes 33 percent of the deaths, but by age 100 it is responsible for only 4 percent, largely because many cancer sufferers die much younger, sometimes from cardiovascular diseases. Prostate cancer, for example, is common among men in their 70s, but many die not from it but from heart

Table 4–6
Age Category in Which Cause-Specific Death Rate of Blacks Drops Below That of Whites, by Sex, 1985–87

Cause	Male	Female
All causes	85+	85+
Diseases of heart	85+	85+
Ischemic heart disease	65–69	80–84
Acute and subacute myocardial infarction	55–59	75–79
Angina pectoris	80–84	80–84
Old myocardial infarction and other chronic ischemic	70–74	80–84
Other diseases of endocardium	70–74	75–79
Malignant neoplasms	...	85+
Digestive organs and peritoneum	85+	...
Respiratory and intrathoracic organs	85+	65–69
Breast	...	65–69
Urinary organs	50–54	60–64
Leukemia	60–64	65–69
Other lymphatic and hematopoietic	80–84	75–79
Cerebrovascular diseases	85+	85+
Intracerebral and other intracranial hemorrhage	85+	80–84
Cerebral thrombosis	85+	85+
Cerebral embolism	70–74	70–74
Other	85+	85+
Diabetes mellitus	85+	85+
Accidents and adverse effects	85+	85+
Motor vehicle	75–79	60–64
Other	85+	85+
Chronic obstructive pulmonary diseases	60–64	55–59
Pneumonia and influenza	85+	75–79
Chronic liver disease and cirrhosis	70–74	70–74

Source: National Center for Health Statistics, Vital Statistics of the United States, 1985, 1986, 1987, Vol. II, Mortality, Part A (Washington, DC: Government Printing Office, 1988–90), Sec. 1, various tables.

disease, stroke, pneumonia, or other causes. Moreover, the declining relative importance of cancer has to be due to the increasing relative importance of other causes because the death rate for cancer among the elderly actually rose 6 percent between 1978 and 1987. Cerebrovascular diseases cause a growing percentage of deaths among all race and sex groups in the ages 65–94, but then a decreasing percentage. The death rate for stroke among persons 65 and over, however, fell from 662 per 100,000 in 1978 to 435 in 1987. As a result of these changes in the three major causes, all other causes collectively rise in relative importance with age, from 27 percent of the deaths in the ages 65–69 to 35 percent among centenarians, as persons become increasingly afflicted by multiple illnesses.

GEOGRAPHIC DISTRIBUTION OF CAUSES

Mapping death rates by cause for states and other geographic units presents certain challenges, such as the need to compute age-adjusted death rates to compensate for variable age profiles. Age standardization is especially important in analyzing the geographic distribution of such causes as heart disease and cancer. Since they are primarily diseases of the elderly, any state with a sizable in-migration of elderly persons, such as Florida and Arizona, will have an above-average death rate for these diseases if only crude rates are used.[49] Mapping cause of death in units smaller than the state, and even in some states, is difficult because of small numbers—either base populations or deaths—in the specific age categories used to compute age-adjusted death rates. The effort is complicated further by attempting cross-classification according to race and sex because the numbers become even smaller.[50]

Given the problems, this section shows 1988 variations by state in the age-adjusted death rates for the two principal causes of death—heart disease and cancer—using the total population of each state with cross-classification by sex but not race. The reader will recognize the practical necessity but also the shortcomings of this procedure, since some variations by state are due to the variable distribution of races and ethnic groups with very different rates for many causes of death. A comprehensive ecological study of cause of death, however, is a major project in itself and beyond the scope of this book.[51]

Heart Disease

The distribution of age-adjusted death rates for heart disease among males and females is shown in Figure 4–2, although different scales are used for the sexes because in every state the death rate among males is close to twice that among females (see also Table 4–7, which gives sex mortality ratios for heart disease).

Among males the death rates for heart disease range from 135 per 100,000 in Hawaii to 278 in Mississippi. Moreover, the rates are generally highest throughout the South (except in Florida), partly because of higher average poverty rates and poorer health care in some sections, and partly because of diet. The rates are also high in several northwestern and midwestern states whose economies are depressed because of declining "rust-belt" industries, such as those in Michigan and Ohio. Several of these states and others with high death rates for heart disease also contain relatively large percentages of blacks, among whom death rates for this cause are much higher than among whites. New York joins the list partly for this reason. Death rates are lower, but still above the national average, in other states with large industrial metropolises, such as New Jersey,

Figure 4–1
Percent of All Deaths for Three Leading Causes Among the Elderly, by Age, 1986–88

PERCENT

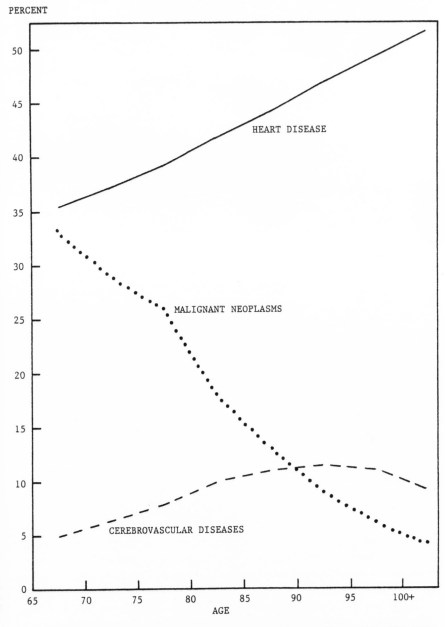

Source: National Center for Health Statistics, *Vital Statistics of the United States*, 1986, 1987, 1988, Vol. II, *Mortality*, Part B (Washington, DC: Government Printing Office, 1988–90), Sec. 8, Table 5.

Figure 4–2
Age-Adjusted Death Rate for Heart Disease, by State and Sex, 1988

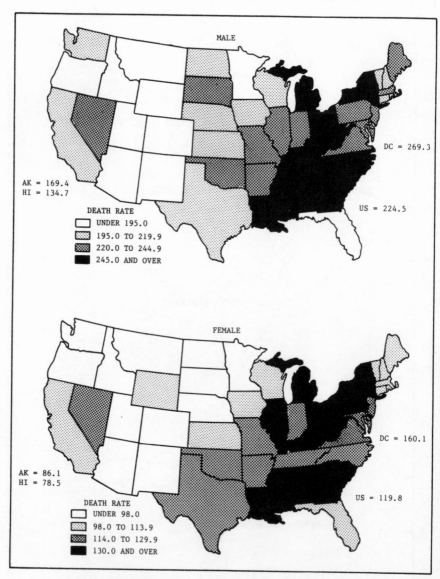

Source: U.S. Bureau of the Census, "State Population and Household Estimates, With Age, Sex, and Components of Change: 1981–88," *Current Population Reports*, Series P–25, No. 1044 (Washington, DC: Government Printing Office, 1989), Table 8; National Center for Health Statistics, *Vital Statistics of the United States, 1988*, Vol. II, *Mortality*, Part B (Washington, DC: Government Printing Office, 1990), Sec. 8, Table 6.

Table 4–7
Male/Female Ratio of Age-Adjusted Death Rates for Heart Disease and Cancer,
by State, 1988

State	Heart Disease	Cancer	State	Heart Disease	Cancer
United States	187.4	146.0	Virginia	190.6	159.4
			West Virginia	183.8	145.1
Northeast			North Carolina	206.8	163.4
Maine	203.9	141.1	South Carolina	178.2	165.7
New Hampshire	200.8	136.9	Georgia	186.0	176.4
Vermont	185.2	134.6	Florida	182.0	146.8
Massachusetts	199.9	138.5	Kentucky	198.3	163.5
Rhode Island	197.5	147.1	Tennessee	191.7	167.9
Connecticut	185.6	140.4	Alabama	187.2	168.0
New York	181.4	141.7	Mississippi	192.1	178.5
New Jersey	185.5	141.5	Arkansas	196.4	157.8
Pennsylvania	183.0	146.0	Louisiana	181.7	161.1
			Oklahoma	190.5	156.9
Midwest			Texas	187.2	152.8
Ohio	185.0	144.8			
Indiana	199.4	147.6	West		
Illinois	182.6	147.4	Montana	197.3	133.5
Michigan	181.1	143.1	Idaho	198.6	132.7
Wisconsin	201.0	144.7	Wyoming	157.0	124.1
Minnesota	208.9	136.6	Colorado	189.1	132.4
Iowa	203.2	145.9	New Mexico	186.6	129.4
Missouri	190.2	158.4	Arizona	194.8	148.0
North Dakota	208.0	147.6	Utah	180.5	129.2
South Dakota	233.1	133.6	Nevada	190.2	138.0
Nebraska	202.1	137.2	Washington	200.0	126.5
Kansas	186.3	151.1	Oregon	199.1	126.8
			California	178.9	132.4
South			Alaska	196.7	145.0
Delaware	172.3	131.8	Hawaii	171.6	142.6
Maryland	178.8	145.9			
District of					
Columbia	168.2	169.4			

Source: U.S. Bureau of the Census, "State Population and Household Estimates, With Age, Sex, and Components of Change: 1981–88," Current Population Reports, Series P-25, No. 1044 (Washington, DC: Government Printing Office, 1989), Table 8; National Center for Health Statistics, Vital Statistics of the United States, 1988, Vol. II, Mortality, Part B (Washington, DC: Government Printing Office, 1990), Sec. 8, Table 6.

Pennsylvania, Indiana, and Illinois. Missouri, Oklahoma, and Arkansas also have fairly high rates, partly because of rural poverty and partly because of their racial composition.

On the other hand, death rates for heart disease among males are lowest in the West (except in Nevada) and in Minnesota and Florida. In the case of Nevada, 82 percent of the population live in the Las Vegas and Reno MSAs, and it could simply be that a large share of persons in those places do not have particularly healthy lifestyles. In the case of Florida, the state's proportion of elderly persons is so large that it has a higher median age than any other state.[52] In 1988 men 65 and over were 10 percent of the nation's male population but 16 percent of that in Florida. Many elderly have migrated from elsewhere, and while not all are affluent and some are poor, neither do they represent a cross-section of the nation's elderly male population. That is, migration selects for elderly persons who are relatively well off financially and who have better access to medical care, private health insurance, and other advantages than do elderly persons in many other places. This is partly because the Florida group is largely a white population living not in inner cities or poverty-stricken rural sections, but rather under circumstances that allow above-average prevention and treatment of heart disease. Furthermore, the migration itself selects persons in relatively good health, often leaving behind those with the most serious heart problems and contributing conditions. Therefore, while the age-adjusted death rate compensates for Florida's age profile, the health advantages of many elderly Florida residents also help lower the state's death rate for heart disease. This also applies to Arizona, although men 65 and over are a smaller share (12 percent) of its male population than in Florida.

Among females the geographic variations in age-adjusted death rates for heart disease range from 79 per 100,000 in Hawaii to 160 in the District of Columbia. The distribution patterns are much like those among men, with some variations. Women do not fare as uniformly badly as men in several southern states, although women's relative advantage in Florida is somewhat less than that of men. Elderly women there are 20 percent of the female population, compared with 14 percent in the nation as a whole, but many are widows and widowhood tends to reduce income significantly, thereby restricting access to medical care for some, despite Medicare, Medicaid, and other publicly sponsored programs. The problem of living alone, discussed earlier, also has an impact.

Women in the upper Midwest also fare well compared with women in many other parts of the country, while men there compare less favorably with men in several other sections. Generally, however, the states with high, medium, or low rates of death for heart disease in one sex repeat the pattern in the other. The antecedents, which affect both sexes, include proportions of minority group members, localized lifestyles (e.g., eating habits, smoking patterns), quality of state and local health care, percentages of persons in poverty, and many others. Therefore, as in the case of infant mortality, high death rates for heart disease are at least partly a local problem that may not always respond well to national efforts applied uniformly in each state and locale, although federal funding is indispensable to fight heart disease.

In the nation as a whole the age-adjusted death rate for heart disease is 87 percent higher among men than women. In 45 states the age-adjusted death rate among men is at least 80 percent higher than among women. In the others it is at least 57 percent higher; they are Delaware, Maryland, the District of Columbia, South Carolina, Wyoming, and California (see Table 4–7). In ten states the rates among men are more than twice those among women, and while the sex mortality gap tends to be widest in the states with the highest death rates for heart disease, there is no clear-cut pattern. Some states with relatively low death rates among both sexes have a wide sex gap, as do others with relatively high death rates among both sexes. This too reflects the important effect of local conditions on sex differentials in death rates for heart disease and suggests the need for locally tailored efforts to reduce both.

Malignant Neoplasms (Cancer)

The state-by-state distribution of age-adjusted death rates for cancer among males and females appears in Figure 4–3, which also uses different scales for the sexes. In the male population the death rates for cancer range from 112 in Utah to 240 in the District of Columbia. Furthermore, the ecological distribution of death rates for cancer are fairly similar to those for heart disease; they are generally highest in the South, again except in Florida for reasons suggested earlier. Cancer death rates are also relatively high in parts of New England and in a broad band of states that stretches west from New York and New Jersey to Illinois and south to include Missouri, Arkansas, and Oklahoma, although the cancer death toll among males in these states does not rival the toll in the Southeast.

The lowest death rates for cancer, again duplicating much of the heart disease pattern, are in the West (except in Nevada); rates are also relatively low in a number of midwestern states. The cancer death rate is especially low in Utah, partly because members of The Church of Jesus Christ of Latter-Day Saints (Mormons) have very low rates of smoking and alcohol consumption. Mormons are 73 percent of all church members in Utah, so their prescribed temperance has a considerable impact on the death rate for the whole state, not just for cancer, but for heart disease and other causes.

Among females geographic variations in age-adjusted death rates for cancer run the gamut from 85 per 100,000 in Hawaii to 142 in the District of Columbia, although these extremes are not as far apart as those among males. The distribution of cancer death rates among women is also quite different from that among men and from the distribution of heart disease among women. Thus, women's highest cancer death rates are in New England, New Jersey, Delaware, West Virginia, and Ohio, along with three states in the West. Intermediate though comparatively high levels appear in parts of the upper South, in several states around the Great Lakes, and in Montana, Washington, and California. Conversely, while the lowest rates show up in parts of the West, those in large parts of the South are not excessively high either. Such states as North and South Carolina, Tennessee, Georgia, and Mississippi have much lower death rates for cancer among women than men; nor do women in the Deep South have the uniformly high death rates for cancer that they have for heart disease.

Cancer is also a more nearly equal killer of both sexes than is heart disease,

Figure 4–3
Age-Adjusted Death Rate for Cancer, by State and Sex, 1988

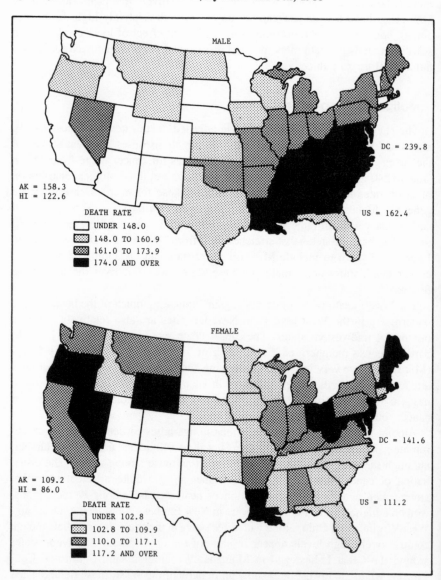

Source: U.S. Bureau of the Census, "State Population and Household Estimates, With Age, Sex, and Components of Change: 1981–88," *Current Population Reports*, Series P–25, No. 1044 (Washington, DC: Government Printing Office, 1989), Table 8; National Center for Health Statistics, *Vital Statistics of the United States, 1988*, Vol. II, *Mortality*, Part B (Washington, DC: Government Printing Office, 1990), Sec. 8, Table 6.

and in every state the sex mortality ratio for cancer is significantly lower than the one for heart disease (see Table 4–7). That does not apply in the District of Columbia, however, where the sex mortality ratio for cancer is slightly above that for heart disease, largely because cancer is so serious in the District of Columbia's large black population. Ratios are smaller elsewhere, partly because progress in lowering the death rates for certain kinds of cancer has been slower than progress in reducing the death rate for heart disease, but largely because death rates for some malignancies have risen, especially among women. Therefore, the differences among the states in sex mortality ratios for cancer are smaller than state-to-state variations in death rates for that cause. In 1988, for example, the highest age-adjusted death rate for cancer among males was 115 percent greater than the lowest and the highest rate among females was 65 percent above the lowest, but the extremes of the sex mortality ratio for cancer differed by only 44 percent. Thus, in most states relatively high death rates for cancer among males are coupled with relatively high rates among females, with some exceptions noted earlier. This situation too reflects the localized nature of high-risk or low-risk conditions for both sexes; it calls for the same localized focus on prevention and treatment suggested for heart disease. Greater federal funding for cancer prevention might help, although the nation does spend about $2 billion annually to prevent and treat malignancies and conduct cancer research, so increased costs would have to be weighed carefully against increased benefits.

NOTES

1. A. H. Pollard, Farhat Yusuf, and G. N. Pollard, *Demographic Techniques*, 3d ed. (Sydney, Australia: Pergamon Press, 1990), p. 64.

2. National Center for Health Statistics, Jack M. Guralnick, Andrea Z. LaCroix, Donald F. Everett, and Mary Grace Kovar, "Aging in the Eighties: The Prevalence of Comorbidity and Its Association with Disability," *Advance Data from Vital and Health Statistics* 170 (Washington, DC: Government Printing Office, May 1989), p. 1.

3. See Michael R. Alderson, *Mortality, Morbidity and Health Statistics* (New York: Stockton Press, 1988), pp. 41–45; Kenneth G. Manton and George C. Myers, "Recent Trends in Multiple-Caused Mortality 1968 to 1982: Age and Cohort Components," *Population Research and Policy Review* 6:2 (1987), pp. 161–176; and Kenneth G. Manton and Eric Stallard, *Recent Trends in Mortality Analysis* (New York: Academic Press, 1984), Chap. 3.

4. National Center for Health Statistics, *Vital Statistics of the United States, 1987*, Vol. II, *Mortality*, Part A (Washington, DC: Government Printing Office, 1990), Sec. 7, p. 8.

5. Ibid.

6. NCHS, Guralnick et al., "Aging in the Eighties," p. 3.

7. National Center for Health Statistics, Gloria Kapantais and Eve Powell-Griner, "Characteristic of Persons Dying of Diseases of Heart," *Advance Data from Vital and Health Statistics* 172 (Washington, DC: Government Printing Office, August 1989), p. 26.

8. Robert D. Retherford, *The Changing Sex Differential in Mortality* (Westport, CT: Greenwood Press, 1975), p. 15. A compilation of studies on the matter is Millicent W.

Higgins and Russell V. Luepker, *Trends in Coronary Heart Disease Mortality: The Influence of Medical Care* (New York: Oxford University Press, 1988).

9. National Center for Health Statistics, Eve Powell-Griner, "Characteristics of Persons Dying from Cerebrovascular Diseases," *Advance Data from Vital and Health Statistics* 180 (Washington, DC: Government Printing Office, February 1990), p. 1.

10. Ibid., pp. 1, 3.

11. Ibid., pp. 5–6.

12. Metropolitan Life Insurance Company, "Deaths from Chronic Obstructive Pulmonary Disease in the United States, 1987," *Statistical Bulletin* 71:3 (July-September 1990), pp. 20–26.

13. National Center for Health Statistics, Thomas F. Drury and Anita L. Powell, "Prevalence of Known Diabetes Among Black Americans," *Advance Data from Vital and Health Statistics* 130 (Washington, DC: Government Printing Office, July 1987), p. 1.

14. Manton and Stallard, *Mortality Analysis*, p. 58.

15. National Center for Health Statistics, Thomas F. Drury and Anita L. Powell, "Prevalance, Impact, and Demography of Known Diabetes in the United States," *Advance Data from Vital and Health Statistics* 114 (Washington, DC: Government Printing Office, February 1986), pp. 1, 8.

16. Rosemary Gartner, "The Victims of Homicide: A Temporal and Cross-National Comparison," *American Sociological Review* 55:1 (February 1990), p. 99. See also Martin Daly and Margo Wilson, *Homicide* (Hawthorne, NY: Aldine de Gruyter, 1988).

17. National Center for Health Statistics, Joseph E. Fitti and Marcie Cynamon, "AIDS Knowledge and Attitudes for April-June 1990," *Advance Data from Vital and Health Statistics* 195 (Washington, DC: Government Printing Office, December 1990), p. 1.

18. See National Center for Health Statistics, Gloria Kapantais and Eve Powell-Griner, "Characteristics of Persons Dying From AIDS," *Advance Data from Vital and Health Statistics* 173 (Washington, DC: Government Printing Office, August 1989).

19. International Planned Parenthood Federation, Tony Klouda, " 'Marginalization'— A Useful Concept for AIDS Projects," *AIDS Watch* 13 (1991), p. 2, from Robert Chambers, *Rural Development: Putting the Last First* (White Plains, NY: Longman, 1983).

20. John E. Anderson, Laura Kann, Deborah Holtzman, Susan Arday, Ben Truman, and Lloyd Kolbe, "HIV/AIDS Knowledge and Sexual Behavior among High School Students," *Family Planning Perspectives* 22:6 (November/December 1990), p. 252.

21. Ibid., pp. 252–253.

22. Ibid., p. 255.

23. National Center for Health Statistics, Deborah A. Dawson and Ann M. Hardy, "AIDS Knowledge and Attitudes of Black Americans," *Advance Data from Vital and Health Statistics* 165 (Washington, DC: Government Printing Office, March 1989), p. 2.

24. Ibid.

25. Ibid., p. 3.

26. National Center for Health Statistics, Deborah A. Dawson and Ann M. Hardy, "AIDS Knowledge and Attitudes of Hispanic Americans," *Advance Data from Vital and Health Statistics* 166 (Washington, DC: Government Printing Office, April 1989), pp. 2–3. See also Barbara V. Martin and Gerardo Marin, "Hispanics and AIDS," special issue of *Hispanic Journal of Behavioral Sciences* 12:2 (May 1990).

27. For the data on variations by race, sex, and behavioral patterns, see National

Center for Health Statistics, *Health, United States, 1990* (Washington, DC: Government Printing Office, 1991), Table 46.

28. National Center for Health Statistics, "Advance Report of Final Mortality Statistics, 1988," *Monthly Vital Statistics Report* 39:7 (Washington, DC: Government Printing Office, November 1990), p. 46.

29. For the data, see National Center for Health Statistics, *Health, United States, 1989* (Washington, DC: Government Printing Office, 1990), Table 23. See also NCHS, Kapantais and Powell-Griner, "Persons Dying from AIDS."

30. NCHS, Drury and Powell, "Diabetes Among Black Americans," p. 1.

31. NCHS, "Advance Report of Final Mortality Statistics, 1988," p. 9.

32. A commentary on Hispanic health levels is Antonia C. Novello, Paul H. Wise, and D. V. Kleinman, "Hispanic Health: Time for Data, Time for Action," *Journal of the American Medical Association* 265:2 (January 9, 1991), pp. 253–255.

33. Ira Rosenwaike, "Introduction," in *Mortality of Hispanic Populations*, edited by Ira Rosenwaike (Westport, CT: Greenwood Press, 1991), pp. 7–8.

34. Ibid., p. 8.

35. Ibid.

36. Ibid., p. 9.

37. Ibid., p. 11.

38. Ibid., p. 10.

39. NCHS, *Health, 1989*, p. 14.

40. Kim Streatfield, Masri Singarimbun, and Ian Diamond, "Maternal Education and Immunization," *Demography* 27:3 (August 1990), p. 447.

41. Kenneth Hill and Anne R. Pebley, "Child Mortality in the Developing World," *Population and Development Review* 15:4 (December 1989), p. 681.

42. NCHS, *Health, 1989*, p. 15.

43. Ibid., p. 17.

44. See Madelyn S. Gould, Sylvan Wallenstein, Marjorie B. Kleinman, Patrick O'Carroll, and James Mercy, "Suicide Clusters: An Examination of Age-Specific Effects," *American Journal of Public Health* 80:2 (February 1990), pp. 211–212.

45. National Center for Health Statistics, Beulah K. Cypress, "Health Care of Adolescents by Office-Based Physicians: National Ambulatory Medical Care Survey, 1980–81," *Advance Data from Vital and Health Statistics* 99 (Washington, DC: Government Printing Office, September 1984), p. 1.

46. Jerald G. Bachman and Lloyd D. Johnston, "Patterns of Drug Use," *Institute for Social Research Newsletter* 15:2/3 (Fall/Winter 1987–88), p. 3.

47. NCHS, *Health, 1989*, p. 21.

48. For the supporting data, see ibid., p. 174. See also National Center for Health Statistics, Thomas F. Drury and Ildy I. Shannon, "Health Practices and Perceptions of U.S. Adults with Noninsulin-Dependent Diabetes: Data From the 1985 National Health Interview Survey of Health Promotion and Disease Prevention," *Advance Data from Vital and Health Statistics* 141 (Washington, DC: Government Printing Office, September 1987).

49. See a Louisiana study to compensate for this problem by Theodore F. Thurmon, Susonne A. Ursin, and Kathleen R. Wiley, "Association of Lung Cancer Death Rates by Parish with Migration Rate by Age Group," *American Journal of Human Biology* 1:6 (1989), pp. 771–784.

50. For a method to map cause-of-death rates in local areas with small populations, see Kenneth G. Manton, Eric Stallard, Max A. Woodbury, Wilson B. Riggan, John P.

Creason, and Thomas J. Mason, "Statistically Adjusted Estimates of Geographic Mortality Profiles," *Journal of the National Cancer Institute* 78:5 (May 1987), pp. 805–815.

51. For a comprehensive effort to map age-adjusted death rates of white males and females in state economic areas, see National Center for Health Statistics, Linda W. Pickle, Thomas J. Mason, Neil Howard, Robert Hoover, and Joseph S. Fraumeni, *Atlas of U.S. Cancer Mortality Among Whites: 1950–1980* (Washington, DC: Government Printing Office, 1987).

52. U.S. Bureau of the Census, "Population Profile of the United States: 1989," *Current Population Reports*, Series P–23, No. 159 (Washington, DC: Government Printing Office, 1989), p. 9.

5 Trends in Cause of Death

Death rates for specific causes decreased significantly as the United States left the early stages of the epidemiologic transition and moved into the last stage. Table 5–1, which gives age-adjusted death rates for various causes in the entire population, and Tables 5–2 and 5–3 which do so for whites and blacks, respectively, show the progress between 1940 and 1988 in virtually eliminating some causes and greatly reducing the virulence of most others, though notably not all.

There are some methodological concerns in comparing the two years because five revisions of the *International Classification of Diseases* altered the categories of causes. For example, some subcategories of cardiovascular diseases were created and others redefined substantially after 1940. The Ninth Revision of the *ICD*, prepared in 1975 and first applied to 1979 data, is used extensively in this book, although the NCHS also modified parts of that revision for U.S. purposes. In addition, while the standard population for computing age-adjusted death rates was enumerated in 1940 and it is best to trace as many trends as possible from that year, some sections also follow detailed trends from later years, often 1950 or even 1979, particularly for subcategories of cancer. In a few cases trends are even shorter. The first cases of HIV infections, for example, were not reported officially until 1981 and a separate category was not created for HIV until 1987.

Despite changes in categories of causes and in the time span used to identify trends, the causes listed in Tables 5–1 through 5–3 are reasonably comparable for 1940 and 1988. The data show many significant reductions and some major increases during nearly a half century of progress in medicine, sanitation, nutrition, and health awareness and maintenance.[1] The changes parallel trends in most other developed countries, where cardiovascular diseases now head the list of causes, followed by cancer and external causes, especially motor vehicle accidents and suicide.[2]

Table 5-1

Age-Adjusted Death Rate for Selected Causes, by Sex, 1940 and 1988

Cause[a]	1940			1988		
	Both Sexes	Male	Female	Both Sexes	Male	Female
All causes	1,076.1	1,213.0	938.9	535.5	696.7	404.4
Cardiovascular diseases	485.8	548.9	423.0	206.6	270.8	155.4
Malignant neoplasms	120.3	115.5	125.7	132.7	162.4	111.2
Digestive organs	55.3	60.7	49.9	30.0	38.6	23.3
Genital organs	23.5	15.3	32.2	13.0	15.6	12.2
Breast	11.8	0.3	23.3	12.6	0.2	23.1
Respiratory organs	7.2	11.1	3.4	39.9	59.7	24.4
Urinary organs	5.8	7.5	4.0	4.9	7.5	3.0
Leukemia	3.9	4.6	3.3	4.8	6.2	3.7
Accidents	72.4	102.4	43.3	35.0	51.5	19.1
Other	46.2	62.0	31.3	15.3	23.3	7.9
Motor vehicle	26.2	40.4	12.0	19.7	28.3	11.3
Pneumonia and influenza	70.2	78.1	62.2	14.2	18.9	11.0
Tuberculosis	45.9	54.1	37.3	0.5	0.8	0.3
Diabetes mellitus	26.6	20.3	33.0	10.1	10.5	9.8
Syphilis	14.4	20.7	8.0	0.0	0.0	0.0
Suicide	14.3	21.9	6.8	11.4	18.7	4.7
Appendicitis	9.9	12.1	7.6	0.1	0.2	0.1
Hernia and intestinal obstruction	9.0	9.7	8.4	1.1	1.1	1.1
Ulcer of stomach and duodenum	6.8	11.1	2.4	1.4	1.8	1.0
Homicide	6.3	10.1	2.5	9.0	13.9	4.2
All other causes	194.2	208.1	178.7	113.4	146.1	86.5

Source: Robert D. Grove and Alice M. Hetzel, Vital Statistics Rates in the United States, 1940-1960 (Washington, DC: Government Printing Office, 1968), Table 62; National Center for Health Statistics, "Advance Report of Final Mortality Statistics, 1988," Monthly Vital Statistics Report 39:7 (Washington, DC: Government Printing Office, November 1990), Table 12.

[a]Deaths per 100,000 population in specified group.

INFECTIOUS AND PARASITIC DISEASES

Data for the death-registration states in 1900 show that infectious and parasitic diseases, then called "general diseases," caused 27 percent of all deaths (26 percent if influenza is excluded). Pneumonia and influenza headed the list, followed by tuberculosis, diarrhea, enteritis, and ulceration of the intestines. Heart disease ranked only fourth, followed by stroke, nephritis, and accidents, and by cancer in eighth place. Senility and diphtheria completed the list of ten leading causes.[3]

Many earlier contagious scourges, including some that caused major epidemics, are nearly gone, but that did not happen until well into the century. Thus,

Table 5-2
Age-Adjusted Death Rate for Selected Causes Among Whites, by Sex, 1940
and 1988

Cause[a]	1940			1988		
	Both Sexes	Male	Female	Both Sexes	Male	Female
All causes	1,017.2	1,155.1	879.0	509.8	664.3	384.4
Cardiovascular diseases	468.9	535.3	402.6	199.2	264.2	147.2
Malignant neoplasms	121.3	117.7	125.5	130.0	157.6	110.1
Digestive organs	56.1	61.6	50.7	28.6	36.6	22.4
Genital organs	22.8	15.5	30.5	12.3	14.5	11.9
Breast	12.0	0.3	23.6	12.5	0.2	23.0
Respiratory organs	7.5	11.5	3.4	39.4	58.0	24.8
Urinary organs	5.9	7.7	4.1	5.0	7.7	2.9
Leukemia	4.1	4.7	3.4	4.9	6.3	3.8
Accidents	71.8	100.1	42.2	34.1	49.9	18.8
Other	45.6	59.9	30.7	14.1	21.4	7.2
Motor vehicle	26.2	40.2	12.2	20.0	28.5	11.6
Pneumonia and influenza	63.0	70.1	55.7	13.6	18.0	10.7
Tuberculosis	36.1	44.0	27.9	0.3	0.5	0.2
Diabetes mellitus	26.7	20.6	32.8	9.0	9.6	8.4
Suicide	15.2	23.1	7.2	12.2	19.8	5.1
Appendicitis	9.7	12.0	7.4	0.1	0.1	0.1
Syphilis	9.7	14.4	4.9	b	b	b
Hernia and intestinal obstruction	8.6	9.3	7.9	1.0	1.0	1.0
Ulcer of stomach and duodenum	6.7	11.1	2.2	1.4	1.7	1.0
Homicide	3.2	5.0	1.3	5.3	7.7	2.8
All other causes	176.3	192.4	161.4	103.6	134.2	79.0

Source: Robert D. Grove and Alice M. Hetzel, Vital Statistics Rates in the United States, 1940-1960 (Washington, DC: Government Printing Office, 1968), Table 62; National Center for Health Statistics, "Advance Report of Final Mortality Statistics, 1988," Monthly Vital Statistics Report 39:7 (Washington, DC: Government Printing Office, November 1990), Table 12.

[a]Deaths per 100,000 population in specified group.

infectious and parasitic diseases, including influenza, still caused 8 percent of all deaths in 1940, but only 2.3 percent in 1988 (1.5 percent if HIV infections are excluded). In 1940 death rates for tuberculosis and syphilis were higher than for several kinds of cancer, homicide, and other causes that now account for a much larger proportion of all deaths. In that year tuberculosis killed over 60,000 people, syphilis more than 10,000, and the two caused 5 percent of all deaths. They were especially virulent among blacks, and 11 percent of their deaths were attributable to those causes, compared with 4 percent among whites. By 1988 effective prevention and treatment had cut the toll to less than 2,000 deaths from

Table 5-3
Age-Adjusted Death Rate for Selected Causes Among Blacks, by Sex, 1940
and 1988

Cause[a]	1940			1988		
	Both Sexes	Male	Female	Both Sexes	Male	Female
All causes	1,634.7	1,764.4	1,504.7	788.8	1,037.8	593.1
Cardiovascular diseases	669.1	691.3	647.1	293.0	360.8	241.1
Pneumonia and influenza	138.1	153.6	122.4	19.6	28.0	13.4
Tuberculosis	132.9	148.0	116.8	2.1	3.2	1.3
Malignant neoplasms	101.5	83.7	119.5	171.3	227.0	131.2
Digestive organs	41.8	46.8	36.4	43.7	58.4	32.5
Genital organs	30.8	13.2	48.8	21.6	30.6	16.4
Breast .	9.2	0.3	18.4	15.3	0.4	27.0
Respiratory organs	4.3	6.3	2.3	49.8	83.4	24.6
Urinary organs	3.7	4.4	2.9	4.8	6.6	3.5
Leukemia	2.1	2.7	1.6	4.5	5.8	3.5
Accidents	81.4	120.3	43.4	43.7	69.0	22.2
Other	55.8	78.8	33.4	25.0	39.4	12.9
Motor vehicle	25.6	41.5	10.0	18.7	29.6	9.2
Syphilis	61.6	85.0	37.3	0.1	0.2	0.1
Homicide	34.2	57.1	12.6	34.1	58.2	12.7
Diabetes mellitus	23.3	15.0	32.2	21.2	19.8	22.1
Hernia and intestinal obstruction	12.8	13.8	11.8	1.9	2.1	1.6
Appendicitis	10.8	12.6	9.0	0.3	0.4	0.2
Ulcer of stomach and duodenum	7.6	11.2	4.0	1.7	2.4	1.1
Suicide	5.1	7.9	2.2	6.8	11.8	2.4
All other causes	356.3	364.9	346.4	193.0	254.9	143.7

Source: Robert D. Grove and Alice M. Hetzel, Vital Statistics Rates in the
United States, 1940-1960 (Washington, DC: Government Printing Office,
1968), Table 62; National Center for Health Statistics, "Advance Report of
Final Mortality Statistics, 1988," Monthly Vital Statistics Report 39:7
(Washington, DC: Government Printing Office, November 1990), Table 12.

[a]Deaths per 100,000 population in specified group.

tuberculosis and only 85 from syphilis, although blacks were still disproportionately affected.

In 1940 at least 500 persons also died from each of several other infectious
and parasitic diseases, including typhoid fever, cerebrospinal meningitis, scarlet
fever, whooping cough, diphtheria, tetanus, septicemia, dysentery, malaria, lo-
comotor ataxia (loss of coordination of movement and balance), measles, poliomyelitis, and acute infectious encephalitis. Some even died from plague,
undulant fever, anthrax, leprosy, relapsing fever, smallpox, and typhus. In 1988

there were 278 deaths from meningitis, 16 from tetanus, and four or less from each of the other causes, with one notable exception—septicemia.

Septicemia has become a more dangerous infectious illness, responsible for nearly 21,000 deaths in 1988 and for 45 percent of all deaths from infectious and parasitic diseases (71 percent if HIV infections are excluded). By comparison, in 1940 septicemia caused only 12 percent of the deaths from infectious and parasitic illnesses. It is primarily a disease of the elderly, often because a weakened immune system cannot resist infection of the urinary tract or of untreated cuts, burns, and other injuries. Thus, in 1988 while persons 65 and over were a little more than 12 percent of the population, they had 65 percent of the deaths from septicemia. The disease has become more serious, therefore, partly because the number and percentage of elderly people grew substantially after 1940, when they were only 7 percent of the total, although the age-adjusted death rate for this cause also rose substantially. Septicemia is much more a disease of blacks than whites, and at all ages its death rate among black males is two to seven times higher than among white males; the greatest discrepancy is in the ages 30–49. The age-specific death rates for septicemia among females are two to five times higher for blacks than whites.

Influenza is now classified separately from infectious and parasitic diseases, but in 1940 it was not and it obviously is contagious. The death rates for influenza vary considerably from year to year as epidemics come and go, but in 1940 it was responsible for a death rate seven times higher than in 1988. In 1940 it killed over 20,000 persons and followed tuberculosis as a leading cause of death from communicable diseases. By 1988 the others were so well controlled that influenza took a larger proportional toll than any other, although it caused less than 2,000 deaths in a population almost twice that of 1940. Still, it is a significant cause among the elderly, since 92 percent of all deaths from influenza occur among persons 65 and over. In 1940 it winnowed the population more evenly because only 45 percent of the deaths were among the elderly, even though the death rate for influenza among persons 65 and over was 100 per 100,000, compared with 59 in 1988.

Human immunodeficiency virus (HIV) infections could have existed much earlier than 1981, but that is when the first few cases were officially reported in the United States.[4] Most persons with HIV infections develop acquired immune deficiency syndrome (AIDS), which is immune system collapse because of infection by one of the human immunodeficiency viruses. At this writing AIDS is a terminal disease.[5] According to the Committee on AIDS Research of the National Research Council, data on AIDS as a cause of death greatly underestimate the actual prevalence of existing cases and the incidence of new cases, partly because of the long latency period without symptoms between contracting HIV and developing AIDS. Moreover, while the lives of persons with HIV are shortened significantly by the infections, some individuals do not have enough symptoms to be detected by the AIDS reporting system, so the incidence of HIV is substantially underestimated.[6] Detecting AIDS in women is especially problematic because of a tendency to focus on the disease among men and to overlook some rather different symptoms in women. The Centers for Disease Control (CDC) reported 31,001 diagnosed AIDS cases in 1988, which represents a rapid increase from 199 cases in 1981, and they reported a total of 83,991 cases for 1981–88.[7] Furthermore, the CDC reported 17,119 deaths from AIDS in 1988, up from 1,416 in 1983, for a total of 53,671 deaths during that period.[8] According to the Committee on AIDS Research, however, there are actually about 1 million cases of HIV, with a possible range of 0.5 to 2 million.[9] Even the higher figure may be too conservative.

These reporting problems and the short time for which data are available make it difficult to measure the trend for deaths from HIV/AIDS; HIV is particularly difficult to pinpoint statistically. It is clear, however, that the trend line is moving rapidly upward and that the actual numbers of AIDS cases and deaths are far above the reported numbers. The disease has become an epidemic in the sense that the number of new cases is growing rapidly, even though it may decline at times and among certain groups (e.g., men who have sex with men), so we are "at the base of a rapidly rising curve of AIDS cases and deaths."[10] Between 1981 and 1990, for example, there were about 101,000 deaths from HIV complications, but more than 31,000 of those occurred in 1990 alone. The rise is so significant for men aged 25–44, that HIV infections reversed the long-term decline in their death rate.[11] It is now the second leading cause of death for that group, surpassing heart disease, cancer, homicide, and all other causes except accidents. This concentration of the disease, however, should not lead us to neglect the needs of older persons with AIDS, and while they are a relatively small share of all cases, they too need more attention.[12]

Even if new cases of AIDS have peaked in some groups, they continue increasing rapidly in others (e.g., heterosexual teenagers). Moreover, even if HIV infections begin to decline, new AIDS cases will develop for several years because of earlier HIV infections, and even if the number of new AIDS cases drops, new HIV infections would not necessarily do so.[13] HIV infections pervade certain groups and the virus itself spreads rapidly through the body and mutates easily, so new infections can be expected to develop, whether at epidemic rates or more slowly. Persons who will continue to be most at risk are intravenous drug users, men who have sex with men, and their partners of either sex and their children. Blacks and Hispanics also have more than their proportional share of AIDS cases, especially those involving intravenous drug use, heterosexual contact, and transmission from mothers to infants; the disease is likely to continue spreading rapidly in those groups. More effective efforts to slow the spread of HIV, based on the modification of risky sexual behavior and intravenous drug use, will help prevent many AIDS cases, even though a preventive vaccine or cure may be far off. Research is needed especially on "sexuality outside of marriage; sexuality with persons of the same gender; sexuality with persons of both genders; sexual contacts for pay; and variations in sexual techniques among various types of sexual partnerings."[14] It is a large order indeed.

CARDIOVASCULAR DISEASES

Age-adjusted death rates for cardiovascular diseases collectively are much lower than a few decades ago. Between 1940 and 1988, for example, the rate fell 64 percent among white females, 63 percent among black females, 51 percent among white males, and 48 percent among black males (see Tables 5–1, 5–2, 5–3). The changes increased the relative advantage of black and white women over men in their respective races, while the advantage of whites over blacks also grew. Thus, longstanding differentials increased despite the substantial drop in death rates for cardiovascular diseases in all race and sex groups. The discrepancy between white males and black females did diminish substantially, but black men, long the most likely to die of cardiovascular diseases, became increasingly worse off relative to all other groups.

The category "cardiovascular diseases" contains a large number of specific causes. The major one—diseases of the heart—includes several subcategories, but ischemic heart disease is by far the most important, although its relative significance is declining. It is commonly manifested as heart attack because of thrombosis (blockage) by a thrombus (blood clot). In turn, ischemic heart disease is caused by acute myocardial infarction, old myocardial infarction, and other specific conditions. Angina pectoris is listed as one of them, though it is not a cause but the pain that occurs when the oxygen supply to the heart is reduced, generally because of artery blockage or high blood pressure.

The second most important form of cardiovascular diseases—cerebrovascular disease (stroke)—is attributable to intracerebral and other intracranial hemorrhage, cerebral thrombosis, cerebral embolism, and other conditions.

Two types of trends show how well various groups fared as death rates for these more detailed causes changed. One trend is the changes between 1950 and 1988 in the importance of the overall category of diseases of the heart and cerebrovascular diseases as causes of death, using the age-adjusted death rate. The second is the changes in the subcategories of those causes from 1979 to the late 1980s, using age-specific death rates. The latter show recent progress in controlling these many specific causes, using comparable categories from the Ninth Revision of the *ICD*. Moreover, changes in age-adjusted rates by race and sex provide less insight into detailed causes than do changes in age-specific rates because cardiovascular diseases diminished more as a cause of death for some groups than for others. In fact, at certain ages the death rates for these causes actually rose.

Diseases of the Heart

The major killer of Americans has changed in several ways. Principally, the long decline in the death rate for heart disease continued at an accelerating pace in the population as a whole and it came to account for a decreasing share of all deaths (see Table 5–4). In the 1950s, for example, the age-adjusted death

Table 5-4
Age-Adjusted Death Rate for Heart Disease, by Race and Sex, 1950-88

Race and Sex[a]	1950	1960	1970	1980	1988	Percent Change 1950-88
All races	307.2	286.2	253.6	202.0	166.3	-45.8
White						
Male	381.1	375.4	347.6	277.5	220.5	-42.1
Female	223.6	197.1	167.8	134.6	114.2	-48.9
Black						
Male	415.5	381.2	375.9	327.3	286.2	-31.1
Female	349.5	292.6	251.7	201.1	181.1	-48.2

Source: National Center for Health Statistics, Vital Statistics of the United States, Vol. II, Mortality, Part A, for data years 1950-87 (Washington, DC: Government Printing Office); NCHS, "Advance Report of Final Mortality Statistics, 1988," Monthly Vital Statistics Report 39:7 (Washington, DC: Government Printing Office, November 1990), Table 12.

[a]Deaths per 100,000 population in specified group.

rate for heart disease fell by an annual average of 0.88 percent, but in the 1960s it dropped an average of 1.14 percent; in the 1970s, 2.03 percent; and in the 1980s, 2.21 percent. In 1950 heart disease caused 37 percent of all deaths, compared with 32 percent in 1988. The pace of decline also accelerated steadily for white and black males, while among females of both races, especially blacks, it accelerated until the early 1980s and then slowed. Between 1982 and 1988, for instance, the death rate among males fell 2.6 percent annually, but among females it dropped only 1.5 percent. This is one reason life expectancy of women at age 65 dropped from 18.8 years in 1982 to 18.6 years in 1984 and remained there, even while the life expectancy of men continued to rise slowly.[15] Even so, in 1988 heart disease accounted for a smaller percentage of all deaths in each age, race, and sex group than in 1950. This killer is coming increasingly under control, though at a slowing pace in some groups.

That control is apparent in the changing death rates for various forms of heart disease in specific age groups, shown in Table 5-5. It gives the percentages by which the death rates for the major types of heart disease changed between 1979 and 1987 in each ten-year age group from 35-44 to 85 and over. The table does not include persons under 35 because the number of deaths from heart disease is relatively small, though some observations about heart disease among infants do appear later.

Ischemic Heart Disease. Ischemic heart disease remains the major form, but the age-adjusted death rate and most age-specific death rates for it are declining. As a result, while ischemic heart disease caused 75 percent of all deaths from

Table 5–5
Percent Change in Age-Specific Death Rate for Type of Heart Disease, by Race and Sex, 1979–87

Race and Sex	35–44 Years	45–54 Years	55–64 Years	65–74 Years	75–84 Years	85+ Years
White Male						
All types heart disease	−23.3	−27.7	−22.8	−20.1	−18.6	−19.8
Ischemic	−32.2	−34.7	−30.3	−28.0	−25.6	−11.6
Acute myocardial infarction	−45.6	−42.7	−40.0	−36.7	−29.7	−12.6
Old myocardial infarction	0.0	−13.8	−7.2	−12.0	−20.9	−11.0
Hypertensive	−7.1	−14.0	−11.9	−17.0	−18.9	−6.8
White female						
All types heart disease	−22.9	−18.0	−14.4	−13.6	−19.3	+1.5
Ischemic	−28.3	−24.8	−22.5	−21.7	−26.8	−9.3
Acute myocardial infarction	−36.5	−31.7	−30.5	−24.6	−26.2	−9.2
Old myocardial infarction	−7.4	−7.6	−4.9	−13.6	−22.4	−9.0
Hypertensive	0.0	−28.6	−20.7	−27.3	−29.6	−6.1
Black male						
All types heart disease	−10.2	−12.6	−16.6	−3.7	−17.1	+33.7
Ischemic	−26.5	−25.4	−28.5	−15.0	−27.0	+16.3
Acute myocardial infarction	−37.3	−33.0	−35.9	−19.9	−28.4	+17.0
Old myocardial infarction	−5.8	−16.3	−15.0	+6.9	−25.3	+15.5
Hypertensive	−19.1	−11.8	−10.1	−13.2	−30.8	+25.7
Black female						
All types heart disease	−23.1	−19.7	−13.0	−4.1	−20.9	+49.7
Ischemic	−35.6	−30.6	−22.3	−13.9	−30.8	+34.2
Acute myocardial infarction	−46.0	−33.6	−28.5	−12.5	−26.4	+35.3
Old myocardial infarction	−18.1	−24.0	−11.6	−15.4	−34.0	+33.8
Hypertensive	−29.8	−24.8	−26.3	−13.7	−31.9	+27.6

Source: National Center for Health Statistics, Vital Statistics of the United States, 1979, Vol. II, Mortality, Part A (Washington, DC: Government Printing Office, 1984), Sec. 1, Tables 7, 9; NCHS, Vital Statistics of the United States, 1987, Vol. II, Mortality, Part A (Washington, DC: Government Printing Office, 1990), Sec. 1, Table 10.

heart disease in 1979, by 1987 it caused only 66 percent. In fact, the decline began among men in most parts of the nation between 1960 and 1965, although the onset of decline varied more widely among women.[16] Furthermore, progress in reducing the death rate from ischemic heart disease is due to greatly improved control over acute myocardial infarction because Americans are achieving better prevention of first heart attacks and higher survival rates among those who have them. The age-adjusted death rate for old myocardial infarction and other chronic heart disease also has fallen, but more slowly than for the acute form. We have learned the importance of immediate treatment for a heart attack in order to prevent death, minimize heart damage, and reduce the probability of a second attack. Measures such as rapid responses to 911 calls, sophisticated cardiac intensive-care units, wider knowledge of cardiopulmonary resuscitation (CPR), and medication to dissolve blood clots have helped create this major health improvement. So have coronary artery bypass surgery, coronary arteriography to show the extent of narrowing in coronary arteries, and other techniques.

Several lifestyle indicators show that preventive measures also are helping reduce the death rate from heart attack. The proportion of persons with high blood pressure, for example, has decreased significantly. The reduction is largely among persons 55 and over, and their percentages with borderline high blood pressure (systolic pressure at least 140, diastolic pressure at least 90) and with elevated blood pressure (160/95 or more) have fallen among blacks and whites of both sexes. Apparently, efforts to create awareness of the connection between high blood pressure and heart disease and stroke are paying off, although the incidence of high blood pressure in several age groups under 55 has risen and requires even more attention to hypertension and its complications. Blacks, to whom a number of publicity efforts have been directed, made particular progress in lowering the incidence of high blood pressure in most ages 20–74, especially 65–74.[17]

The incidence of cigarette smoking also has fallen sharply among males of both major races, especially those aged 18–24, and since it contributes to so many deaths from several causes, any significant reduction has an important impact on the death rates for those causes.[18] In 1965, for example, 53 percent of white men aged 18–24 and 63 percent of the blacks were smokers, but by 1987 the proportions had fallen to 30 percent and 25 percent, respectively. This is encouraging progress among young black men, but the incidence of smoking remains higher among older black than white men, and in all ages above 25 the improvement was slower for blacks. Therefore, about 40 percent of all black men and 30 percent of all white men still smoke cigarettes. Women show a different pattern: blacks aged 18–24 have reduced their incidence of smoking the most, followed by whites aged 35–44, but in every age group and for both races the incidence of cigarette smoking fell more slowly among women than men. It even rose 42 percent among white women aged 65 and over and 65 percent among blacks in those ages; it also rose 11 percent among black women aged 45–64.[19] This slower progress in "kicking the habit" helps explain why the death rate for ischemic heart disease fell less for elderly white women than elderly white men and why the death rate among blacks rose more for elderly women than elderly men, although other factors also are involved.

One such factor is the level of serum cholesterol, which is classified as high at 240 milligrams or more per deciliter (borderline high serum cholesterol is 200–239). The percentages of white males and females with high serum cholesterol fell in all age groups 20–74; the percentage of black males with the condition dropped only in the ages 35 and over, whereas among black women the proportion dropped in the ages 25 and over. Furthermore, some evidence suggests that cholesterol levels no higher than 240 may be acceptable for men without manifestations of cardiovascular disease and that the goal of 200 or less might better be limited to those with such manifestations.[20] If so, part of the reduction in death rates for heart disease, stroke, and some other causes could be due to many persons having achieved an acceptable cholesterol level between 200 and 240. Furthermore, cross-national studies show considerable variation in the association between coronary heart disease and diets high in fat and cholesterol.[21] Heart disease has many precipitating conditions, including genetic programming that causes individuals to be more or less susceptible to adverse effects of a high-fat diet and other risk factors. Furthermore, efforts to change dietary habits seem more successful on a person-by-person basis than as mass campaigns to alter lifestyle, especially in a heterogeneous population with a plethora of food preferences, many reinforced by ethnic heritage.[22] Nevertheless, at present an overall goal of serum cholesterol of 200 or below seems highly desirable.

Another factor implicated in ischemic heart disease is obesity (body mass of 27.8 kilograms or more per square meter among males and 27.3 kilograms or more among females). In this case, progress is not so impressive. Among white males aged 20–34 the percentage of overweight persons fell, but it rose in the ages 35 and over. Among white females it fell in the ages 55–74 but rose in the ages 20–54. Among black males the incidence of obesity also dropped in the ages 20–34 but rose in the older years, while among black females it fell only in the ages 35–44 and rose in all others. Thus, to the extent obesity is associated with the probability of ischemic heart disease, all groups still have much room to improve.[23]

Heart Failure. Heart failure produces the second largest number of deaths from heart disease, but only about 8 percent as many as the ischemic form. Almost 86 percent of heart failure deaths are due to congestive heart failure, in which the pumping action of the heart falters, though it does not stop as in the case of cardiac arrest. Congestive heart failure usually results from a weakened heart due to disease or a mechanical problem in the heart valves; one may also suffer left-ventricular or right-ventricular heart failure. The term "congestive" applies when the entire heart is affected because the inefficient pumping action either allows blood to accumulate in the veins that carry blood from the lungs, thus allowing them to become congested with fluid, or it enables blood to accumulate in the veins that lead to the heart from other parts of the body so that fluid collects in those other parts, especially the legs.[24]

Heart failure became a proportionately more important form of heart disease in the 1980s. It caused 3.4 percent of all heart disease deaths in 1979 but 5.4 percent in 1987, partly because the death rate for ischemic heart disease fell faster, and partly because of aging of the population. As expected, heart failure is overwhelmingly a disease of the elderly and becoming more so; in 1979, 86 percent of the deaths occurred among persons 65 and over, compared with 91 percent in 1987. Moreover, while heart failure is often not fatal until long after its onset, in persons 85 and over it is a degenerative disease that reflects wearing out toward the upper end of the life span. Heart failure is often accompanied by the breakdown of other organs as well, and it can be either an underlying or accompanying cause of death.

Hypertensive Heart Disease. Hypertensive heart disease produces the third largest number of deaths from heart disease, but only 4 percent as many as the ischemic form. Moreover, death rates have fallen more slowly for hypertensive than ischemic heart disease. Hypertensive heart disease is especially prevalent among blacks, with a death rate from 1.6 to 13.5 times greater than that for whites, depending on age and sex. The discrepancy is particularly great in the ages 25–64, though it does diminish in the older years. Given the large racial differential, any significant change in the death rate among blacks for this cause has a greater impact on their overall losses from heart disease than is true among whites.

Increases in Death Rates for Heart Disease. Death rates for all forms of heart disease have risen significantly for blacks aged 85 and over, thus tending to reduce the mortality crossover effect, although the data for these ages should be treated cautiously (see Table 5–5). The increase results partly from the cumulative effects of long-term conditions among blacks, such as higher rates of hypertension, obesity, and high serum cholesterol. The increase has an economic component too because very elderly blacks are much more likely than whites to be poor, to progress out of poverty more slowly, to perceive medical attention as an unaffordable luxury, and to have less access to medical care, partly because of changes in the Medicaid system in the 1980s and partly because of such practical problems as lack of transportation. The elderly of both races have lower poverty rates now than in 1979, but from then until 1988 the poverty rate fell 25 percent among elderly whites, but only 11 percent among elderly blacks. Furthermore, while the change represents steady improvement for whites, the poverty rate actually rose among older blacks in the late 1980s.[25] In addition, in 1989, with all government transfer payments and Medicaid figured in, 13 percent of the white population aged 75 and over but 30 percent of the black group fell below the poverty line.[26]

A higher poverty rate among elderly blacks is not the only reason for the increase in their death rate for heart disease, but it does mean they have less access to intensive care in coronary units and to other death controls and are, therefore, more likely to die of heart disease. In one sense, the sophisticated ways of preventing and treating heart disease, especially heart attack, have passed them by. At the same time, death rates for heart disease among blacks have fallen in the ages up to 85, so the oldest people are most vulnerable. Larger proportions of persons who might have died at younger ages earlier in the century now remain alive longer, which may mean that the blacks who now survive to 85 are not as sturdy as their aged predecessors. In turn, current losses from heart disease may reflect the inclusion of more vulnerable people who manage to make it to 85 but who have poor prospects for continued survival.

A demographic factor is at work in the increases as well, stemming from the aging of the population. Not only are the numbers and percentages of persons

85 and over growing, but they are increasing faster than any other age group (e.g., 30 percent between 1980 and 1988).[27] That by itself does not explain the increase in their death rates for heart disease, but within the category 85 and over the very old are increasing fastest; the group 100 and over, for instance, increased 260 percent between 1980 and 1988. The greater concentration of persons at the upper end of the 85-and-over group does help explain the higher death rates in some race and sex categories, since the oldest persons are most likely to die of heart disease. Consequently, while adverse socioeconomic conditions and other disadvantages that retard health care raise the death rate for heart disease among some groups, especially black males, the pattern of aging also helps explain the increase. As the large baby boom cohort ages and the proportion of persons at highest risk of heart disease grows, the death rate for that cause will rise unless risk factors diminish further and various types of therapy improve.[28]

Finally, women 85 and over are especially likely to live alone and to lack another person, particularly a spouse, to help maintain a healthful diet and other healthful habits, to summon emergency aid, to provide companionship, and to do other things that directly or indirectly affect health and mortality. When living alone reflects weak social networks, persons are less likely to remain healthy and even survive than those with strong networks of family, friends, and neighbors.[29] In 1989, for example, 51 percent of the women aged 65–74 lived with a spouse while 34 percent lived alone. In the ages 75–84, however, the proportion living alone rose to 51 percent, and in the ages 85 and over to 54 percent. Furthermore, 56 percent of white women aged 85 and over but only 43 percent of blacks lived alone, which is one reason the racial mortality crossover occurs in the older years. These figures represent no change for whites since 1979, but an increase in the proportion of elderly black women living alone may help reduce or even eliminate the crossover at some point. The percentage of elderly white males living alone also remained unchanged while that of black men rose, although men of either race are much less likely than women to live alone.[30] To the extent this increase in solitary living aggravates poverty, neglect, stress, anxiety, and other conditions that predispose one to heart disease, it too helps explain the significant increase in the death rate for that cause among the nation's oldest blacks.

Death rates for heart disease also increased between 1979 and 1987 among white male and female infants, though not among blacks. The number of deaths from that cause is small—less than 1,000 in 1987—but increases of 24 percent for white males and 42 percent for white females are of concern. The forms of heart disease that kill infants, however, are quite different from those that claim adults. In 1987, for example, while ischemic heart disease caused 67 percent of all deaths from heart disease, it accounted for less than 3 percent among infants, though there are cases. Congenital anomalies of all kinds caused 21 percent of infant deaths and heart defects were the leading form, accounting for 30 percent of the deaths. Furthermore, as the infant death rate for infectious and parasitic diseases fell, congenital anomalies, including heart problems, caused a larger share of all infant deaths, although the death rate for congenital anomalies also rose for other reasons, mostly an increase in the percentages of underweight and immature births to unmarried and very young women.[31]

Cerebrovascular Diseases

The age-adjusted death rate for stroke decreased even more sharply between 1950 and 1988 than the rate for heart disease, though for many of the same reasons. The greatest decrease was among black women, but in all race and sex groups the death rate for stroke fell at least 60 percent (see Table 5–6). Furthermore, in the 1950s, 1960s, and 1970s these reductions accelerated for whites and blacks of both sexes, and even though the rate of decline slowed somewhat in the 1980s for all except white males, it did continue. The black/white variation, however, changed differently for males and females. In 1950, for example, the death rate for stroke among black females was 95 percent greater than that among white females, but by 1988 the difference was 83 percent. The difference for males, however, increased from 68 percent to 93 percent, and black males have a far higher stroke death rate than any other group, partly because of a potentially lethal combination of hypertension and stress. The changes represent some progress and some regress in controlling conditions that contribute to stroke, and they show the continuing health and survival disadvantage of black males.

Strokes are far less common in children than in the elderly, of course, but as many as 7 percent of black children who have sickle-cell anemia have strokes severe enough to cause paralysis. Moreover, since sickle-cell anemia occurs in about one of every 400 black infants, the danger of stroke is significant. In fact, it is the most common side effect of the anemic condition due to narrowing of arteries in the brain, arterial blockage, and death of parts of the brain, although the nature of arterial narrowing is different than in adults and the cause is unknown. Now, however, medical personnel can identify high-risk children by using ultrasound to measure the speed of blood flow through the arteries and by inferring the degree of narrowing. In turn, it is easier to know which children need preventive measures, especially blood transfusions, and the technique should serve as a screening tool on a large scale, especially since it is relatively inexpensive.[32] The challenge is to make it universally available to black children with the anemia.

Much of the progress in reducing the death rate for stroke in adults took place before 1980, and in most specific age groups the reductions have been slower since then. Unlike the case with heart disease, however, there have been no reversals for any race or sex at any age, though year-to-year fluctuations do raise the rate occasionally and temporarily. In addition, the principal improvement occurred among people from their mid-40s to 65 or so regardless of race and sex, though this is not to minimize the substantial progress among all groups in reducing the death rate for stroke. Much of the improvement is due to the same factors that lowered death rates for ischemic heart disease: better treatment and

Table 5-6
Age-Adjusted Death Rate for Cerebrovascular Diseases, by Race and Sex,
1950-88

Race and Sex[a]	1950	1960	1970	1980	1988	Percent Change 1950-88
All races	88.6	79.7	66.3	40.8	29.7	-66.5
White						
Male	87.0	80.3	68.8	41.9	30.0	-65.5
Female	79.7	68.7	56.2	35.2	25.5	-68.0
Black						
Male	146.2	141.2	122.5	77.5	57.8	-60.5
Female	155.6	139.5	107.9	61.7	46.6	-70.1

Source: National Center for Health Statistics, Vital Statistics of the
United States, Vol. II, Mortality, Part A, for data years 1950-88
(Washington, DC: Government Printing Office); NCHS, "Advance Report of
Final Mortality Statistics, 1988," Monthly Vital Statistics Report 39:7
(Washington, DC: Government Printing Office, November 1990), Table 12.

[a]Deaths per 100,000 population in specified group.

rehabilitation of stroke victims, more effective prevention of second and third
attacks, attention to high blood pressure, reduced incidence of high serum cholesterol, and less cigarette smoking in several age groups. In turn, these and
other measures diminish the incidence of arteriosclerosis (hardening of the arteries), thrombosis, embolism (blockage by a loose blood clot), aneurysm (bal-
looning of the artery wall), and other conditions that cause stroke. Many
aneurysms are congenital, though they may also be caused by bacterial infections,
atherosclerosis, and high blood pressure, and some can be diagnosed and cor-
rected before they burst and cause a fatal or disabling stroke. The surgery is
risky, however, and the reduction in death rates for stroke due to aneurysm
depends largely on prevention by slowing atherosclerosis and controlling blood
pressure.

MALIGNANT NEOPLASMS (CANCER)

Changes in death rates for various forms of cancer are far less a success story
than changes for heart disease, stroke, and most other major causes. Unlike the
case for heart disease and stroke, the crude death rate for cancer has climbed
steadily among blacks and whites of both sexes. That is partly due to aging of
the population, but between 1950 and 1980 even the age-adjusted rate rose 20
percent among white males and 80 percent among black males, though it did

Table 5-7
Age-Adjusted Death Rate for Malignant Neoplasms, by Race and Sex, 1950-88

Race and Sex[a]	1950	1960	1970	1980	1988	Percent Change 1950-88
All races	125.3	125.8	129.8	132.8	132.7	+5.9
White						
Male	130.9	141.6	154.3	160.5	157.6	+20.4
Female	119.4	109.5	107.6	107.7	110.1	-9.3
Black						
Male	126.1	158.5	198.0	229.9	227.0	+80.0
Female	131.9	127.8	123.5	129.7	131.2	-0.5

Source: National Center for Health Statistics, Vital Statistics of the United States, Vol. II, Mortality, Part A, for data years 1950-87 (Washington, DC: Government Printing Office); NCHS, "Advance Report of Final Mortality Statistics, 1988," Monthly Vital Statistics Report 39:7 (Washington, DC: Government Printing Office, November 1990), Table 12.

[a]Deaths per 100,000 population in specified group.

¶ fall slightly for both groups after 1980. Among black and white females the age-adjusted rate vacillated so that it was somewhat lower in 1988 than in 1950 among white females and about the same among black females in both years, but in the 1980s it also climbed among females of both races (see Table 5-7). These are net results of reductions in the death rate at some ages for certain types of cancer, combined with increases at other ages for other types. For example, between 1979 and 1987 the death rate for all cancer rose 3 percent among white males in the ages 55-64 and 12 percent among those aged 85 and over, but it decreased in all other age groups from 35-84. Among white females the rate rose in all ages 55 and over, with the largest increase for women aged 65-74. The black male population registered increases in the ages 65-74 and especially 85 and over, but decreases in all other ages. Black females saw increases in the ages 35-44, 55-74, and 85 and over (see Table 5-8). The dramatic rise in the death rate among blacks aged 85 and over and the more modest increase among whites in those ages is partly due to the changing proportional size of the groups that make up the 85-and-over category, discussed earlier. The increase also is due partly to a growing risk for certain kinds of cancer, considered next.

Malignant Neoplasms of the Digestive Organs

There has been some progress in reducing the death rate for cancer of the digestive system, and after 1940 the age-adjusted rate decreased among both sexes (see Table 5-1). Significant racial differences appear in the patterns of

Table 5-8
Percent Change in Age-Specific Death Rate for Type of Malignant Neoplasm, by Race and Sex, 1979-88

Race and Sex	35-44 Years	45-54 Years	55-64 Years	65-74 Years	75-84 Years	85+ Years
White male						
All types malignant neoplasms	-9.2	-11.3	+2.5	-0.4	-1.5	+11.8
Digestive organs	+6.8	+9.1	+17.9	+17.2	-11.1	+23.4
Respiratory organs	-26.1	-14.9	+4.1	+3.2	+8.9	+34.2
Genital organs	-12.5	-6.9	+7.4	+4.7	-3.8	+14.1
Urinary organs	0.0	-12.0	+1.7	-11.7	-11.2	+0.5
Leukemia	0.0	-13.0	-2.8	-7.8	-8.3	-8.3
Other lymphatic malignancies	0.0	-3.7	+6.4	-9.6	+9.3	-28.6
White female						
All types malignant neoplasms	-10.5	-7.7	+2.9	+11.3	+1.4	+10.6
Digestive organs	-16.4	-14.2	-13.2	-9.5	-15.4	-6.4
Respiratory organs	-12.3	+7.0	+35.5	+65.6	+66.5	+51.0
Breast	-4.7	-8.7	+0.7	+8.1	+5.0	+1.7
Genital organs	-18.2	-21.7	-14.5	-4.0	-7.6	+1.7
Urinary organs	-14.3	0.0	-1.1	-3.7	-9.6	+7.2
Leukemia	-16.7	-12.2	-4.4	-0.5	-12.1	+15.1
Other lymphatic malignancies	-5.0	-7.8	+3.1	+14.5	+18.5	+55.3
Black male						
All types malignant neoplasms	-14.9	-6.4	-3.8	+12.0	-0.2	+63.7
Digestive organs	-6.7	-10.4	-10.0	-2.5	-10.5	+47.2
Respiratory organs	-27.1	-6.5	-2.2	+26.6	+17.3	+103.5
Genital organs	-77.8	-16.0	-4.2	+6.5	-4.2	+67.3
Urinary organs	0.0	-24.5	0.0	-7.4	-7.2	+39.0
Leukemia	-23.1	+5.4	-5.4	-4.0	0.0	+46.7
Other lymphatic malignancies	+47.5	+15.6	-15.3	+35.5	+19.6	+81.2
Black female						
All types malignant neoplasms	+2.2	-7.5	+0.7	+12.3	-8.1	+68.2
Digestive organs	-13.6	-18.6	-7.6	-0.7	-16.1	+67.3
Respiratory organs	+2.3	-0.2	+40.6	+89.0	+24.4	+117.9
Breast	+29.0	+4.7	+14.8	+24.8	-5.2	+59.2
Genital organs	-24.5	-19.8	-10.0	-12.2	-10.5	+65.1
Urinary organs	-37.5	+11.4	-16.4	+19.8	-20.1	+64.3
Leukemia	-7.7	-11.5	-2.1	+8.3	-8.9	+41.9
Other lymphatic malignancies	0.0	-3.7	-4.1	+13.9	-5.4	+105.2

Source: National Center for Health Statistics, Vital Statistics of the United States, 1979, Vol. II, Mortality, Part A (Washington, DC: Government Printing Office, 1984), Sec. 1, Tables 7, 9; NCHS, Vital Statistics of the United States, 1987, Vol. II, Mortality, Part A (Washington, DC: Government Printing Office, 1990), Sec. 1, Table 10.

change, and while decreases were especially striking among whites, in the black population the death rate for digestive cancer fell modestly among women but rose significantly among men (see Table 5–2 and 5–3). Therefore, black men have a greater risk for this cause, while black women and whites of both sexes are less susceptible.

Within the category of digestive system cancer, colorectal cancer was the principal form in 1979 and 1988. Its two components changed differently, however, and cancer of the colon now accounts for a larger share of all deaths from digestive cancer, rectal malignancies for a smaller share. These changes occurred for blacks and whites of both sexes, and it seems clear that the incidence of colon cancer is closely associated with the consumption of red meat and animal fat, though not vegetable fat.[33] In fact, the traditional southern diet high in fat, fried meats, and dairy products helps elevate the incidence of colon cancer among that region's relatively large black population, and many whites as well.

Pancreatic malignancies among white men caused a smaller share of deaths in 1988 than 1979, but the relative importance of that cause rose among white females and blacks of both sexes. Cancer of the liver and intrahepatic bile ducts also accounted for a larger proportion of the digestive cancer deaths in 1988 than 1979 among blacks and whites of both sexes. Cancer of the esophagus increased in relative importance among whites of both sexes, but decreased among black males and females. The proportional importance of cancer of the stomach and of the gall bladder decreased among all groups. Thus, several forms of digestive cancer, especially of the colon, have become more significant causes of death, partly because of immoderate use of alcohol and the widespread consumption of high-fat, low-fiber foods, reflected in the incidence of obesity among various groups discussed earlier. Other forms of digestive cancer are now better controlled.

Throughout the 1980s the death rate for digestive cancer increased among white men in all ages over 35, largely because the relative importance of colon cancer rose (see Table 5–8). White men stand alone in these increases for digestive cancer, however, because the death rates fell among white women and among blacks of both sexes, except in the ages 85 and over. In addition to the statistical problems with that age category, many deaths that would have occurred at younger ages in earlier decades now get postponed to the oldest years. Persons who would have died in their 60s or 70s from heart disease, stroke, diabetes, and even cancer survive longer because of better prevention and treatment, but eventually they die of those diseases. Therefore, to some extent the substantial rise in the death rate for all forms of cancer among the oldest persons represents improvements in avoiding early death.

Malignant Neoplasms of the Respiratory Organs

The age-adjusted death rate for cancer of the lung and other respiratory organs rose dramatically after 1940, largely because of cigarette smoking, but also because of other carcinogens. They include such things as certain pesticides and fungicides in agriculture, cotton dust in textile mills, acid fumes in the metal plating industry, factory and automobile emissions, radon gas, and a host of others with effects not fully understood. Whatever the reasons, between 1940 and 1988 the death rate for respiratory cancer rose 454 percent in the population as a whole, 425 percent among white males, 629 percent among white females, 970 percent among black females, and 1,224 percent among black males (see Table 5–1, 5–2, 5–3). Furthermore, while respiratory cancer was responsible for only 6 percent of all cancer deaths in 1940, it caused 30 percent in 1988. Except for HIV/AIDS in the 1980s, this is the sharpest upward trend in the list of causes of death. In both 1979 and 1988 cancer of the trachea (windpipe), bronchus (its two branches), and lung was responsible for at least 94 percent of all respiratory cancer deaths among blacks and whites of both sexes, but in 1987 the percentages were even higher than in 1979. This left cancer of the larynx and other parts of the respiratory system to account for a decreasing percentage.

Table 5–8 shows which age groups had the great rise in the death rate for respiratory cancer. Most of the large increases occurred in the ages 55 and over, though with variations by race and sex. Between 1979 and 1987 the death rate for this cause dropped among white men aged 35–54, and even among older men it did not rise more than 9 percent in any age category until 85 and over, when it increased 34 percent. Among white women the death rate started to rise about age 45 and it continued to do so at an increasing pace until age 85. Thus, middle-aged and elderly white women are major victims of lung cancer. The death rate among black males fell until age 65, after which it increased, but not as fast as among women of both races. Finally, the death rate for lung cancer increased most among black women, beginning at ages 35–44, though the rate remained level for those aged 45–54. It is especially high for women 65–74 and 85 and over. Conversely, the death rate for respiratory cancer decreased in most groups under 35, though their number of deaths is relatively small.

Why the dramatic increase in the death rate for respiratory cancer among middle-aged and elderly persons, especially women? (1) Lung cancer generally takes a long time to develop after a person begins to bathe the lungs with tobacco smoke, so antecedents of the disease may be present for decades before the first symptoms appear or it is detected by X-rays. (2) Women who are now middle-aged or elderly began to smoke in large numbers when norms changed, the taboo against using tobacco faded, and smoking became fashionable and sophisticated. This change accelerated during World War II as the roles of women shifted in concert with the need for paid workers, and it persisted in the postwar decades. (3) Many men also became smokers during World War II, enjoying one of the few luxuries that wartime service allowed, and even when the nation became aware of the dangers of cigarette smoking they were unable or unwilling to quit. (4) Smoking came to be perceived by teenagers as a rite of passage to adulthood, strongly reinforced by their own age group and by intensive advertising, and they acquired a long-term habit that is now reflected in lung cancer among aging adults. (5) Even so-called secondhand smoke, or passive smoking, takes a long time to make its effects felt, although perhaps 17 percent or so of all lung cancer is caused by spending one's childhood and adolescence in a home with at least one regular smoker.[34] (6) Lung cancer caused by other conditions than cigarette smoke also usually requires long exposure to a carcinogen, and the disease then appears among middle-aged and elderly persons.

Since the death rate for lung cancer has dropped among persons in their 30s and even 40s and 50s in some race and sex categories, we can expect rates in the older ages to decline as well. Most people know the dangers of cigarette smoking and the percentage of new smokers in the teenage and young-adult population has fallen. In 1974, for example, 27 percent of males aged 12–17 had smoked during the month they were interviewed, as had 50 percent of those aged 18–25. By 1988 those proportions were down to 12 and 36 percent, respectively. The proportions dropped from 24 to 11 percent among females aged 12–18 and from 47 to 35 percent among those aged 18–25.[35] Furthermore, the incidence of smoking has fallen significantly among blacks and whites, males and females, in all age categories from 18–24 to 65 and over, with two notable exceptions—white women aged 65 and over and black women aged 45 and over.[36] The largest decreases were among persons aged 18–34, and if the percentage of smokers continues to fall, the death rate for lung cancer eventually should drop in all age, race, and sex categories. If the nation continues to reduce other respiratory pollutants, the incidence of lung cancer from them also should drop.

Malignant Neoplasms of the Female Breast

Breast cancer is not exclusive to women, but the number of cases among men is so small (289 deaths in 1988) that this section focuses on women, 41,172 of whom died from the disease in 1988. The trend for breast cancer is not as discouraging as for lung cancer, but even so the age-adjusted death rate did not fall in the female population as a whole. Thus, Tables 5–1, 5–2, and 5–3 show that between 1940 and 1988 all women collectively and white women specifically made virtually no progress in fending off death from breast cancer, while the rate among black women increased significantly. Between 1979 and 1988 the death rate for breast cancer among whites rose in all ages 55 and over and decreased in the younger age categories (see Table 5–8). Among blacks, however, the rate rose in all ages except 75–84, where it changed relatively little. An increase of 29 percent in the ages 35–44 is particularly striking because unlike the case with some other malignancies, the rise in the death rate for breast cancer is not mostly a phenomenon of the elderly. In fact, cancer is the principal cause of death among women in their late 30s and 40s and breast cancer is the worst culprit in those ages. One reason the reported incidence has increased, however, is because more cases are detected early by self-examination and mammography, and widespread diagnosis of the disease will lower the death rate for it. Nevertheless, one woman in nine can now expect to get breast cancer, though partly because greater life expectancy increases the period of susceptibility.

The lack of progress in reducing the age-adjusted death rate for breast cancer among women collectively is the net result of a much worse situation among blacks than whites, though even in the ages where the death rate among whites fell, the declines were modest. Consequently, there is ample room for better early detection to catch the disease before it has metastasized through the lymph system and bloodstream to other organs. Black women especially need such efforts, although poor women of any race are especially unlikely to detect the disease early, partly because many are badly educated about it and partly because they cannot afford or do not understand mammography. For example, 30 percent of black women but only 16 percent of whites have never heard of the procedure, which helps explain why the diagnosed incidence of breast cancer is higher among whites while the death rate for it is higher among blacks. If breast cancer is found early, newer treatment leads to fewer radical mastectomies and more lumpectomies and other conservative procedures. Untreated, however, the disease is almost always fatal, though its progress depends on whether it is the small-cell type, which readily metastasizes, or the large-cell form, which tends to be more localized and easier to treat.

Unfortunately, we do not yet understand much of what causes breast cancer, so deliberate prevention is difficult, although various studies provide promising clues. A high-fat, low-fiber diet, for example, seems strongly implicated in the incidence of breast cancer, just as it is in several other diseases. Menstruation before age 12, menopause after 50, and childbearing after 30 also seem to correlate with the incidence of breast cancer, and those events have become more common. There is also a genetic defect involved that causes the disease to afflict some families much more than others and which may be responsible for 5–10 percent of all breast cancer.[37] Furthermore, at least one effort to prevent other diseases may increase the risk of breast cancer somewhat. Postmenopausal women who take estrogen, especially over a long period, reduce their chances of heart disease, stroke, and osteoporosis and raise their average life expectancy, but they also increase the risk of cancer of the breast and endometrium (lining of the uterus).[38] For most women the benefits of estrogen replacement outweigh the risks, but for those with a family history of breast cancer it may be more prudent to forgo estrogen replacement. At least they should be alert to breast lumps and signs of uterine bleeding. Despite these intriguing leads, however, the link between precipitating factors and breast cancer remains much less clear than the ties between smoking and lung cancer and between exposure to ultraviolet light and skin cancer.[39]

Malignant Neoplasms of the Genital Organs

The trends in age-adjusted death rates for genital cancer reflect progress and regress. In the population as a whole the rate in 1988 was just over half that in

1940, but there was very little change among males, contrasted with a substantial reduction among females (see Table 5–1). In the white population, the death rate for this cause dropped 6 percent among males but 61 percent among females (see Table 5–2). In the black population, however, the death rate for genital cancer rose 132 percent among males and fell 66 percent among females (see Table 5–3). Changes between 1979 and 1987 continued some but not all of the long-term trend, but its magnitude decreased: the death rate rose slightly among black and white men and continued dropping moderately among women of both races. Mortality from cancer of the genital organs is minor among men compared with that among women, and the continuing decreases for women are due to better detection and treatment of ovarian, cervical, and related forms of cancer.[40]

Between 1979 and 1987 the age-specific death rates generally showed decreases for genital cancer until well into the older years, and the losses from this cause increasingly are concentrated among the very old, as are most other forms of cancer. The death rate among white males dropped until ages 55–64, rose briefly, decreased again in the ages 75–84, and rose among persons 85 and over (see Table 5–8). Among black males the rate fell at all ages up to 55–64, rose and fell modestly, and then increased by two-thirds in the 85-and-over category. The rate among females of both races fell in all age groups except 85 and over.

The changes among men are due almost entirely to fluctuations in the relative importance of prostate cancer, the changes among women to variations in cancer of the cervix, other parts of the uterus, and ovaries. Between 1979 and 1987 prostate cancer came to account for an even larger percentage of all deaths from genital malignancies among both black and white men, although the rate is especially high among blacks. The increases are typical of an aging population in which most males who survive to old age develop prostate cancer. It is not nearly as virulent as other cancers, however, and may remain dormant for a long time and show no symptoms apart from those of an enlarged prostate, although it may metastasize, usually to the bone structure. Though in one sense prostate cancer is "normal," it also kills about 30,000 men annually, so it does need to be diagnosed, watched, and treated if necessary, even though elderly men with it are more likely to die of something else. Perhaps the prostate cancer blood test, announced in 1991, will significantly reduce deaths from the disease.

Cancer of the ovaries is still the principal genital malignancy among white women and its relative importance as a cause of death increased between 1979 and 1987. Cancer of the cervix and other parts of the uterus now account for a smaller share of the deaths. Among black women cervical cancer was the principal form of genital cancer in both years, though its relative importance decreased as the proportional significance of cancer of the ovaries and other parts of the uterus than the cervix increased. Even so, with timely detection there should be very few deaths from cervical cancer, since 80–90 percent of those found early can be cured, so the present rate among black women reflects their poorer access to the Pap test and treatment.

Malignant Neoplasms of the Urinary Organs

The age-adjusted death rate for cancer of the urinary organs decreased 25 percent among women between 1940 and 1988, but it remained unchanged among men. While the rate among white women fell, however, that among black women rose, and while the rate among white men remained unchanged, that among black men more than doubled (see Tables 5–1, 5–2, 5–3). Furthermore, in most groups the increases occurred at the oldest ages, although there were also increases for black men aged 45–54 (see Table 5–8). Both in 1979 and 1987 cancer of the kidney was the principal form of urinary cancer among white men until their later 60s, after which bladder cancer took first place, partly as a consequence of complications from an enlarged prostate. Among white women the death rate for bladder cancer surpassed the rate for kidney cancer in the ages 70–74. Among black men that change in ranking occurred after age 65, and among black women after age 55.

Leukemia

The age-adjusted death rate for leukemia rose among males and females of both major races between 1940 and 1988, though by a larger percentage in the black than the white population (see Tables 5–1, 5–2, 5–3). From 1979 to 1987, however, much of the increase stopped. It continued among white males 85 and over and modestly among black males aged 45–54 and black females aged 65–74. Leukemia does not kill at the rates of many other malignancies, but apart from the elderly it still claims a sizable number of children, adolescents, and young adults. In fact, in both 1979 and 1987 it was the leading malignant neoplasm in all ages 0–29. After age 30 the death rate for leukemia continues to rise steadily, but the rates for cancer of the digestive system, breast, and genital system all rise faster and displace leukemia from first place.

Leukemia is actually one type of cancer in the category of malignant neoplasms of the lymphatic and hematopoietic tissues, a residual category not shown in Table 5–8. That overall group of diseases also includes lymphoma (malignant tumor of the lymph glands), one form of which is Hodgkin's disease that is easier to cure than the rare but more life-threatening non-Hodgkin's lymphoma. Between 1979 and 1987 the death rate for lymphatic malignancies other than leukemia decreased among white males in all age groups except 55–64; among white females in the ages under 55; among black males only in the ages 55–64; and among black females in all age groups except 65–74 and 85 and over. In the other age categories the rates rose, but on balance lymphatic cancer has become less important as a cause of death among black females and whites of both sexes, while the age-adjusted rate for it has increased among black males. Furthermore, elderly black men are not the only victims of that increase because lymphatic cancer is the only malignant neoplasm for which the death rate increased in the ages 35–44.

Other Malignant Neoplasms

Three other malignant neoplasms each caused at least 5,000 deaths in 1987: cancer of the lip, oral cavity and pharynx (the part of the throat between the tonsils and the larynx, or voice box); melanoma of the skin; and brain cancer. Among blacks and whites of both sexes the incidence of lip, oral, and pharynx cancer decreased between 1979 and 1987, not just in the total population of all ages, but also in the group 65 and over. Many forms of the disease are easily seen and can be treated early with a high rate of cure, although cancer of the tongue can metastasize much faster than the others.

Melanoma of the skin is the most serious of the three types of skin cancer and it has become a more significant cause of death among whites, though it is a minor one among blacks. In fact, the high melanin level in black skin offers substantial protection against sun damage, which causes long-term changes in many whites who then contract melanoma. It is potentially deadly in a short time, so any increase in its rate is serious. Many more people, however, now know the dangers of exposure to the sun and take precautions that earlier generations did not, thus preventing melanoma from becoming an even more significant cause of death, although the myth of a "healthy" tan and damage to the ozone layer may offset many of those precautions and raise the death rate. Melanoma is largely a disease of middle-aged and elderly people with many years of exposure to ultraviolet rays, and it is increasing in relative importance among them, especially white males. Thus, persons 65 and over accounted for 40 percent of the deaths from melanoma in 1979 but for 47 percent in 1987. In 1979 men aged 65 and over had 36 percent of all melanoma deaths among white males, but in 1987 they had 53 percent. In the white female population the increase was from 44 to 50 percent.

Brain cancer accounted for a larger share of all cancer deaths in 1987 than in 1979 among white males and females, especially those 65 and over, but for a smaller share among blacks of both sexes. Unlike other neoplasms, brain cancer and benign brain tumor are both dangerous because as they grow the skull prevents growth outward and they compress brain tissue, although the malignant form spreads and the benign type generally does not. The elderly have most of these cancers, often because they have metastasized from another site in the body. Therefore, the increase between 1979 and 1987 in the importance of brain cancer as a cause of death among whites 65 and over is due in large part to metastasis from other malignancies; it seems closely related to the substantial increase among the elderly in death rates for respiratory cancer and certain other malignancies discussed earlier.

OTHER DISEASES

Age-adjusted death rates for several other causes listed in Tables 5–1, 5–2, and 5–3 dropped off substantially between 1940 and 1988 and those causes became far less significant. They are diabetes mellitus, pneumonia and influenza, ulcer of the stomach and duodenum, appendicitis, and hernia and intestinal obstruction. In fact, a few are on the list of significant causes only because of their importance in 1940, and by 1988 they had become minimal causes at most ages. Some, however, continue to claim many lives among the most vulnerable persons—the very young and the very old—and diabetes actually had a much higher age-adjusted death rate among black males in 1988 than in 1940. In addition, a few causes not listed in Tables 5–1, 5–2, and 5–3, some added as categories after 1940, remain significant causes that claimed 20,000 or more lives in 1988. They include atherosclerosis; chronic obstructive pulmonary diseases; chronic liver disease and cirrhosis; and nephritis, nephrotic syndrome, and nephrosis.

Diabetes Mellitus

The death rate for diabetes was cut by almost half among males of all races and by over two-thirds among females. In the white population the decreases were about that same magnitude, since whites are a large majority of the total, but in the black population the death rate among females fell by just under a third, while that among males rose by a third. Diabetes is especially likely to go undiagnosed among poor blacks and to be haphazardly treated even if it is known, partly because many poor people do not control their weight or diet in order to manage insulin-dependent diabetes. In fact, the increase in the death rate for diabetes partly reflects the relatively large increase in the incidence of obesity among black men aged 45 and over.

Between 1979 and 1987 progress in reducing the death rate for diabetes was reversed in many age groups among both blacks and whites, and while age-specific death rates did not rise greatly in most cases, in some they did. The overall pattern reflects limited improvements in controlling the disease in recent years. For example, among white males the death rate for diabetes rose in all groups of persons aged 25–64 and 75 and over; among white females it increased in the ages 35–54 and 75 and over. Among black males the rate increased in all age groups 45–85 and over; among black females it did so in the ages 65 and over. The largest increases were in the oldest years, as with several other causes of death, but increases in the 30s, 40s, and 50s were also significant. Since diabetics have a relatively high risk of atherosclerosis and thus of high blood pressure, heart attack, and stroke, any increase in the incidence of diabetes is likely to increase the incidence of those diseases as well and to slow reductions in death rates for them. Diabetes can also cause chronic kidney failure and diabetes retinopathy, in which the blood vessels of the retina constrict and die, leading to temporary or even permanent loss of vision. It is related to other illnesses as well, some of which can result in death. Consequently, while the long-term decrease in the death rate for diabetes in all groups except black males implies progress in lowering the death rates for a number of other causes, the short-term increase is not encouraging in this respect. Nevertheless, diabetes is controlled well enough in many sufferers so that the automobile accident rate, long relatively high because of blackouts related to blood-sugar levels, has dropped close to that of nonsufferers, and many persons with diabetes no longer have restricted driving privileges.[41]

Pneumonia

Pneumonia and influenza are frequently paired in the official data reports, but except for a few epidemic years pneumonia causes over 95 percent of the deaths in that category. We examined death rate trends for influenza earlier and can consider pneumonia separately here. The relative importance of pneumonia has declined sharply since 1940, when it had the third highest age-adjusted death rate, although in 1988 it was still the seventh leading cause (see Table 5–1). The death rate for pneumonia decreased among blacks and whites of both sexes, though more among females than males and blacks than whites. In the changes between 1979 and 1987 the age-specific death rate for pneumonia dropped by half among infants and fell for all other age groups under 25. After 25, however, the death rate rose among white males aged 25–44 and 55 and over and it also increased among white females in the ages 55 and over. Black men aged 25–44 and 65 and over experienced significant increases, as did black women aged 35–44 and 65 and over. Thus, pneumonia grew far less virulent for the population as a whole and for infants, children, adolescents, and young adults, but in several older age groups it became a more significant cause of death after 1979.

Ulcer, Appendicitis, and Hernia and Intestinal Obstruction

The specific diseases in this category have not disappeared as causes of death but have declined sufficiently so as to require little attention, except to show how well death-control techniques can prevent or cure disease and avoid fatalities (see Table 5–1). In this sense, stomach and duodenal ulcer, appendicitis, and hernia and intestinal obstruction have diminished as causes of death nearly as much as the infectious and parasitic diseases of a half century ago.

Tables 5–2 and 5–3 show the extent to which death rates for ulcer and the other two causes have dropped among blacks and whites of both sexes. The rate for ulcer, however, dropped less among white males than white females and blacks of both sexes and less than the death rates for appendicitis—nearly eradicated as a cause of death—and hernia and intestinal obstruction. The last is still fairly important among blacks, but in all race and sex categories its age-adjusted death rates fell at least 85 percent between 1940 and 1988. These causes are heavily concentrated in the oldest population and becoming more so: in 1987 about 82 percent of all deaths from ulcer occurred among persons 65 and over, compared with 74 percent in 1979. The change for appendicitis was from 61 percent to 69 percent, that for hernia and intestinal obstruction from 79 to 85 percent. In part, these causes are more dangerous in the older years because they are less likely to be recognized and treated than in younger people; or they finally reach an acute stage in persons who have learned to live with the chronic stage; or the sufferers cannot afford timely treatment; or surgery for these conditions is more dangerous because their oldest victims are weakened by other illnesses.

Chronic Obstructive Pulmonary Diseases

Bronchitis, emphysema, asthma, and related diseases became more significant causes of death, again in part because of aging of the population, though most age-specific death rates also rose. Between 1979 and 1987 the death rates for these diseases dropped among black and white infants and among somewhat older white children, but not among black children. Among white males aged 5–74 the death rates in 1987 were about the same as or a little lower than in 1979, but in the ages 75 and over they were much higher in 1987. The same is true for white females, except the escalation began about age 45 and was greater than for males in all older ages. Among black males aged 1–44 the death rate for these causes rose between 1979 and 1987, dropped in the ages 45–54, and then increased sharply. Among black females the rate rose in every age group. As with lung cancer and related maladies, the death rates for chronic obstructive pulmonary diseases escalated among adults, especially the elderly, as another consequence of smoking behavior, although other predisposing conditions also are involved. They all contribute to relatively high death rates for several causes that are preventable or curable with more healthful lifestyles and better access to medical care.

Liver Disease and Cirrhosis

Changes in mortality from liver diseases, especially the two viral forms of acute hepatitis and cirrhosis, partly reflect changes in other health-threatening patterns of behavior, in this case the cumulative effects of alcohol abuse. These diseases are almost entirely an adult problem, although a few infants are infected and do die. The death rate for liver problems among whites is commonly ten times as high in the ages 25–34 as 14–24; it jumps sharply in the ages 35–44, doubles in each of the next two age ranges, reaches its peak in the 60s and early 70s, and then declines in the older years, although at every age the death rate among males is at least twice that among females. In the male population death rates are much higher among blacks than whites, but while that racial difference applies among females up to age 65, thereafter it reverses. So, liver diseases disproportionately affect males, especially blacks.

These relationships by race and sex did not change much between 1979 and 1987, but the death rates for liver diseases did drop significantly in virtually all age groups up to 65–74, regardless of race or sex. They increased somewhat among persons 85 and over, except there was no change among the oldest black men. Thus, the several causes of liver disease, such as alcoholism, infections by acute or chronic hepatitis, malnutrition, and congestive heart failure, seem less destructive than a decade earlier, even though the nation still has serious problems with alcoholism in most age groups, including children and teenagers.

Atherosclerosis

Atherosclerosis is a significant cause of death itself, producing 22,000 deaths in 1988, most from unspecified forms of the disease, though some deaths were identified by atherosclerosis of the aorta, renal arteries, and arteries of the extremities. It is more important, however, as a contributing condition to deaths from heart disease and stroke. Deaths from atherosclerosis are heavily concentrated in the older years, with 95 percent occurring after age 65 and 70 percent after 80. The disease became much less important as a cause of death among the elderly, however, for between 1979 and 1987 the death rate for it fell significantly in every age group, regardless of race and sex. It even dropped in the ages under 65, where the number of deaths from this cause is very small. The number of deaths from atherosclerosis has declined since 1979, despite growth of the elderly population. The number of stroke victims also decreased, but the number of deaths from heart disease did not, despite the falling death rate for that cause. Thus, because atherosclerosis now takes a smaller toll than it did, by implication it is better controlled as a contributing cause of heart attack and stroke among elderly people and as a long-term antecedent of those diseases among younger ones. The reduction in deaths from atherosclerosis, therefore, is partly responsible for the reduction across the age spectrum in death rates for heart disease, which claims 35 times as many lives as atherosclerosis itself, and for stroke, which claims seven times as many.

Nephritis and Related Diseases

Nephritis, nephrotic syndrome, and nephrosis are a class of renal, or kidney, diseases. The most common subcategory, which accounts for 92 percent of the total, is kidney failure and disorders resulting from impaired renal function and small kidneys. In turn, kidney failure may be acute, chronic, or end-stage, when dialysis becomes necessary. There are cases of these diseases at all ages, including infants and children with congenital renal problems, especially nephrotic syndrome; in fact, congenital anomalies of the genitourinary tract are more common than those of any other organ system.[42] They are not fatal nearly as often as other congenital anomalies, especially of the heart, but the death rate for kidney diseases is higher among infants than any other age group up to 55, and it is especially severe among black male infants.

Although deaths from nephritis and nephrosis are somewhat more evenly distributed across the age spectrum than deaths from many other causes, 83 percent still occur among persons 65 and over. Between 1979 and 1987 the death rate for these causes among white males fell slightly at most ages, though it rose sharply among men 85 and over. Among white females and black males the rate fell somewhat until ages 55–64, after which it rose. It also rose among black females 65 and over. Thus, there was relatively little progress and even some regress after 1979 in reducing the death rate for these causes. Part of the increase in the older years is due to the increased relative importance of prostate cancer among men and other diseases among women that obstruct the flow of urine and produce acute kidney failure, though a sudden decrease in blood pressure from bleeding, heart attack, or other causes also may induce kidney failure. In the chronic form, high blood pressure, kidney stones, and other conditions can cause kidney inflammation that finally results in kidney failure. Since many of these conditions among the growing elderly population are difficult to control, it is not surprising that kidney ailments remain a relatively serious cause of death.

ENVIRONMENTAL CAUSES

The three principal causes of death that stem not from disease but from dangers in the sociocultural environment—accidents, suicide, and homicide—show different patterns of change in the past half century.

Accidents

The age-adjusted death rate for accidents decreased significantly between 1940 and 1988 among all race and sex groups, but in different ways. For example, the death rate for motor vehicle accidents did not fall nearly as much as that for other types of accidents, regardless of race and sex (see Tables 5–1, 5–2, 5–3). After 1940 the work place became safer; the proportion of workers engaged in dangerous occupations declined; there was a large exodus from farms where the accident rate from falls and encounters with animals and machinery was particularly high. Jobs in construction, mining, certain kinds of manufacturing, and other typically dangerous industries became safer, partly as a result of efforts by companies, labor unions, and the Occupational Safety and Health Administration (OSHA), created in 1970, while a growing share of the work force entered relatively safe service occupations. Other hazards, including many in the home, diminished substantially; better fire prevention was especially important. Thanks to these changes, the death rate for "other accidents" was cut 76 percent among white females, 64 percent among white males, 61 percent among black females, and 50 percent among black males. In this case as with so many other causes of death, black males experienced the least progress.

The age-adjusted death rate for motor vehicle accidents also fell, but not very much among females of both major races. It dropped 29 percent among black and white males, but only 8 percent among black females and 5 percent among white females. The years after 1940 saw greater percentages of women driving automobiles and other motor vehicles, so their statistical risk of fatal accidents rose. Much the same is true of black males as the black middle class increased and more people were able to afford automobiles and as more black workers took jobs in transportation. Furthermore, unprecedented proportions of teenagers, among whom the accident rate is relatively high, became drivers.

The death rate for motor vehicle accidents changed unevenly between 1950 and 1988, with increases for all groups until 1970, followed by decreases. As a result, the age-adjusted death rate for this cause was lower among white males and blacks of both sexes in 1988 than in 1950, though it was higher among white females (see Table 5–9). The vehicular death rates among white and black males are roughly similar and so are those among white and black females, so the principal difference is by sex, not race. In the 1980s the rates also fell for all groups except black females, though their losses from motor vehicle accidents continue to be the smallest among the race and sex groups.

Age-specific death rates for accidents also show changes between 1979 and 1987. During that time the death rate for all accidents decreased among nearly all age groups except persons 85 and over, partly for statistical reasons mentioned earlier and partly because living alone and certain other conditions predispose to accidents among the elderly and may prevent immediate attention. Among males and females of both major races, therefore, death rates for all accidents decreased in each age group 0–84, but they rose for white males and blacks of both sexes aged 85 and over. The same patterns generally apply separately to motor vehicle accidents and to other types, though with a few exceptions. Thus, while death rates for motor vehicle accidents fell for blacks and whites of both sexes in most ages, they rose in the ages 85 and over, except for black males. The rates also increased among black male infants and black females aged 25–44 and 65–74. The death rates for nonvehicular accidents were lower in 1987 than in 1979 in all age, race, and sex groups, again except for persons 85 and over. They are especially susceptible to falls, drowning, fire, inhalation or ingestion of food or other substances, and medical and surgical accidents. Thus, overall progress in preventing death by accident has been good, but infants in some race and sex groups and the very old have not shared in that progress.

Suicide

Self-destruction is less common now than in 1940 among whites of both sexes but more common among blacks of both sexes (see Tables 5–2 and 5–3). The change between 1940 and 1988 was greatest among black males, whose age-adjusted death rate for suicide rose 49 percent, compared with 9 percent among black females. The rate decreased 14 percent among white males and 29 percent

Table 5-9
Age-Adjusted Death Rate for Motor Vehicle Accidents, Suicide, and Homicide, by
Race and Sex, 1950-88

Cause, Race, and Sex[a]	1950	1960	1970	1980	1988	Percent Change 1950-88
Motor vehicle accidents						
White male	35.9	34.0	40.1	34.8	28.5	-20.6
White female	10.6	11.1	14.4	12.3	11.6	+9.4
Black male	39.8	38.2	50.1	32.9	29.6	-25.6
Black female	10.3	10.0	13.8	8.4	9.2	-10.7
Suicide						
White male	18.1	17.5	18.2	18.9	19.8	+9.4
White female	5.3	5.3	7.2	5.7	5.1	-3.8
Black male	7.0	7.8	9.9	11.1	11.8	+68.6
Black female	1.7	1.9	2.9	2.4	2.4	+41.2
Homicide						
White male	3.9	3.9	7.3	10.9	7.7	+97.4
White female	1.4	1.5	2.2	3.2	2.8	+100.0
Black male	51.1	44.9	82.1	71.9	58.2	+13.9
Black female	11.7	11.8	15.0	13.7	12.7	+8.5

Source: National Center for Health Statistics, Vital Statistics of the
United States, Vol. II, Mortality, Part A, for data years 1950-87
(Washington, DC: Government Printing Office); NCHS, "Advance Report of
Final Mortality Statistics, 1988," Monthly Vital Statistics Report 39:7
(Washington, DC: Government Printing Office, November 1990), Table 12.

[a]Deaths per 100,000 population in specified group.

among white females. Nevertheless, the basic relationship persists: white males
are the most likely to kill themselves, followed by black males, white females,
and black females. Females are more likely to attempt but fail at suicide, however, often intentionally by using means that allow time for rescue (e.g., sleeping
pills), contrasted with instantly fatal means used more often by men (e.g.,
firearms).

The sociocultural setting changed significantly between 1940 and 1950 and it
influenced the death rate for suicide. The Great Depression still had a partial
hold on the country in 1940 and serious deprivation by no means had disappeared;
it created enough anomie and other intolerable psychosocial states to influence
significant numbers of persons, especially white males, to commit suicide. By
1950 the depression and World War II were history and the country was relatively
prosperous and stable, so the suicide rate fell during that decade, though more
for whites than blacks (see Table 5-9). In succeeding decades the rates rose for
all groups except white females, though the increase was greatest among black
men.

The rise among blacks, especially males, is partly a result of the breakdown in family structure as a support/constraint network and of a decline in the importance of religion. It is also partly an unintended consequence of the greater incorporation of blacks into the social system and a decrease in their marginality, though not to the point where expectations are fully realized. The civil rights movement, desegregation, improvements in education for blacks, and other changes raised the expectations of blacks and helped replace fatalistic resignation with hopefulness. "Black" was transformed from an epithet into a symbol of race pride and accomplishment, and black leaders dreamed hopefully that blacks would participate fully in the system's opportunities. However, the actual changes did not go far enough and many blacks, especially young men but also many young women and their children and the elderly, found the dream unfulfilled. Fueled partly by constraints imposed by lingering and even resurgent racism, this frustration created a collision between the expectation that norms would work to the full advantage of black people and the realization that they did not. The basic contradiction produced status inconsistency, anomie, alienation, marginality, a sense of failure, and other negative consequences for many blacks. In turn, some used suicide to deal with these intolerable conditions, and the death rate for suicide among black males is higher than ever.[43] Alcohol and drug abuse fit partly into the same context; they are also implicated in many suicides.

Much has been said about rising suicide rates in recent years, but the increases are limited to certain high-risk age groups. In the white male population, for example, the suicide rate rose 100 percent between 1979 and 1987 in the ages 5–14, increased modestly in the ages 14–24, and rose by increasing percentages in the ages 35 through 85 and over, with the largest jump among very elderly men. In the white female population the rate decreased at every age without exception. The suicide rate tripled among black males aged 5–14, rose slightly among those aged 35–44, and increased sharply in all age groups 65 and over, but it fell in the other ages. The rate doubled among black females aged 5–14 but dropped in every other age group. Thus, suicide is still most common among elderly men of both races and the death rate for it is increasing, largely because of mental health problems, especially depression and alcoholism, though loss of a spouse, poor physical health, and other conditions that accompany aging for many males may be contributing factors. The suicide rate for elderly black and white women, already only a fraction of that of men, is dropping. The most spectacular increases in the suicide rate are among children and young adolescents, but the numbers are still small. In 1979, for example, 1,939 persons aged 5–14 killed themselves, and while 2,152 did so in 1987, suicide remains a minor cause of death among children, though this should neither obscure the trend nor discourage efforts to prevent suicide.

Homicide

Homicide and legal intervention increased after 1940 as a cause of death among all race and sex groups, but while the age-adjusted death rate for this cause is much lower among whites than blacks, it rose far more rapidly among whites (see Tables 5–1, 5–2, 5–3). Decade-by-decade increases from 1950 to 1988 show that same pattern, although between 1980 and 1988 the age-adjusted death rate for homicide fell to some extent among males and females of both major races, with the largest decrease among black males (see Table 5–9). Thus, even though homicide is in the leading five causes of death among black males and the leading ten among black females, there was some modest improvement in the 1980s, though with variable changes by age and some reversals in the 1990s. Most of the improvement for all races occurred among adults aged 35–84. Conversely, between 1979 and 1987 the death rate for homicide rose sharply among black and white infants of both sexes, which suggests a significant increase in fatal child abuse, often associated with the misuse of alcohol and drugs. Some of the same applies to increases in the death rate for homicide among black men and women 85 and over, whose rates more than doubled during the 1950–88 period. There were also increases among white females aged 5–14, partly as a result of child abuse, and aged 25–34, partly due to spouse abuse. In addition, some jobs women do, especially in convenience stores, put them at risk for homicide, and if a women is going to die at work, in about 42 percent of the cases she will be murdered, compared with 12 percent of the deaths at work among men.

The death rates for homicide rose among black males aged 5–14 and also 14–24—ages when gang battles, drug-related shootings, and attacks by black men on other black men are most likely. Black females incurred no increases in the death rate for homicide except in infancy and very old age, and infants were the only age group among white males for which the homicide rate rose. Thus, increases in the murder rate are not widely distributed across the age spectrum, but are largely confined to its most vulnerable extremes, some groups of children, and young black men.

The persistence of homicide differentials implies the persistence of other conditions that affect various groups and persons differently. Thus, homicide rates are relatively high when levels of self-esteem and the threshold for violence are low; divorce is also a high-risk factor for homicide among both sexes. In addition, exposure to official violence, sometimes inflicted by police officers, is positively associated with homicide rates, as are weak social integration and the shift of large numbers of women away from traditional roles.[44] Some of these conditions contribute to anomie, in which certain behavioral expectations, such as marital permanence and role predictability, do not work for many persons who then experience stress, anxiety, and anger and who may reject other norms, including those that prohibit murder. In this sense, homicide and suicide have much in common.[45]

Homicide is strongly associated with structural poverty and the perception that it is illegitimate, and persons forced to survive as subordinates in a social system of economic discrimination are especially likely to die of homicide, particularly domestic killings.[46] This fact accounts for a substantial amount of the racial difference in homicide, since "the structuring of economic inequality on the basis of ascribed characteristics [e.g., race] is a particularly important source of lethal violence in contemporary societies."[47] Inasmuch as the personal economic situation of many blacks deteriorated in the 1980s, along with the larger economic setting in which many live, it is hardly surprising that their rates of homicide are many times higher than those among whites. In fact, these conditions affect homicide rates far more than do changes in the proportions of young men, and the United States is actually almost unique in its rising homicide rate as the percentage of young males increased.[48]

Economic and racial discrimination are not the only sources of rising homicide rates among the African-American population, and perhaps not even the primary ones. If they were, homicide rates would have been much higher in the 1950s than at present, since the relative economic status of the black population and discrimination were much worse in the past. Therefore, the factors discussed earlier play at least as important a role as economic and racial discrimination. It is necessary to add, however, that blacks who rank low on the socioeconomic scale have tended to become increasingly isolated into homogeneously poor neighborhoods, partly because those who can afford to get out do so, leaving few black middle- and upper-class role models in those neighborhoods. Some of the high incidence of homicide in those places is related to the poverty, lack of opportunities, frustration, aggression, and violent ways of interacting and even surviving that characterize the neighborhoods. It also is related to family breakdown, the pervasiveness of drugs, the inexhaustible supply of sophisticated weapons, the status attached to violent behavior, and other conditions that pervade many poor, black inner-city areas.

Many conditions just discussed typically have distinguished rural from urban communities and have helped produce significant homicide differentials between them. Some of those differentials persist and many rural populations continue to be more law abiding than urban groups, largely because social bonds are stronger in some rural communities and they produce greater social cohesion, which reduces the incidence of crime, including homicide. Furthermore, the agricultural segment of the rural population, especially persons with stable land tenure arrangements, have the lowest rates of violent crimes.[49] There are several exceptions to these generalizations, however, partly because of variations in the composition and ecological distribution of rural populations. Those in which members are widely dispersed may lack close community cohesion, relying instead on ties in relatively isolated domestic units, where much homicide— rural or urban—takes place.[50] Rural populations with relatively strong community cohesion, on the other hand, continue to have relatively low homicide rates, although such communities are changing significantly. Thus, as divorce rates rise, as more individuals rely on occupational and other ties outside the community, and as religious heterogeneity and other forms of diversity increase, rural areas seem to incur rising rates of homicide and other types of victimization, though on the whole not yet at levels common in urban populations.[51] Thus, the rural advantage in homicide rates persists in many places but has weakened in others.

MALE/FEMALE MORTALITY RATIOS

Preceding sections suggest that death rates for most causes fell faster among females than males and that male/female mortality ratios rose. Thus, according to the 1940 list of 11 major causes of death, by 1988 the sex gap in age-adjusted death rates among whites widened for seven and narrowed for the rest, including motor vehicle accidents and homicide, although decreases for ulcer, appendicitis, and hernia are not very significant because those diseases now produce so few deaths (see Table 5–10). On the other hand, the causes for which the male/female differential grew include the major killers—cardiovascular diseases and malignant neoplasms of the digestive, genital, and urinary organs, along with leukemia. In addition, the sex differential increased for deaths from diabetes mellitus, pneumonia and influenza, nonvehicular accidents, and suicide. The death rates for malignant neoplasms and diabetes among males in 1940 were actually well below those among females, but by 1988 males had no such advantage. On balance, therefore, females not only continue to have lower death rates than males for virtually all major causes, but their advantage has increased for many specific causes. Respiratory cancer, motor vehicle accidents, and homicide stand out as causes of death for which the female advantage has eroded, but by no means has it disappeared: in all three cases the age-adjusted death rates among males are more than twice those among females.

Early in the epidemiologic transition, however, the sex differential was not wide, and women in their childbearing years even had death rates above those of men because of high maternal mortality rates, not just in the United States, but in several other countries for which historical data are available.[52] Apart from deaths among women due to maternal causes and breast cancer and large losses of men to accidents and prostate cancer in pre-industrial societies, however, it appears that death rates of females were not much different than those of males because the major causes of death were much less sex selective than the present major causes.[53] In fact, in the nineteenth century infectious and parasitic diseases and certain other causes took higher tolls of females than males in some age groups, especially 5–19 because of tuberculosis and 25–44 because of pregnancy and childbirth.[54] These conditions are now far less significant causes of death, and the death rate among females is rarely above that among males at any age or for any cause that affects both sexes.

BLACK/WHITE MORTALITY RATIOS

Trends in this chapter also show that while death rates for most causes fell among blacks and whites, the decrease was faster among whites and therefore the black/white mortality ratio generally increased (see Table 5–11). In fact,

Table 5-10
Male/Female Age-Adjusted Mortality Ratio for Cause of Death, by Race, 1940
and 1988

Cause of Death[a]	White		Black	
	1940	1988	1940	1988
All causes	131.4	172.8	117.3	175.0
Tuberculosis	157.7	250.0	126.7	246.2
Malignant neoplasms	93.8	143.1	70.0	173.0
Digestive organs	121.5	163.4	128.6	179.7
Respiratory organs	338.2	233.9	273.9	339.0
Genital organs	50.8	121.9	27.1	186.6
Urinary organs	187.8	265.5	151.7	188.6
Leukemia	138.2	165.8	168.8	165.7
Diabetes mellitus	62.8	114.3	46.6	89.6
Cardiovascular diseases	133.0	179.5	106.8	149.7
Pneumonia and influenza	125.9	168.2	125.5	209.0
Ulcer of stomach and duodenum	504.6	170.0	280.0	218.2
Appendicitis	162.2	100.0	140.0	200.0
Hernia and intestinal obstruction	117.7	100.0	117.0	131.3
Accidents	237.2	265.4	277.2	310.8
Motor vehicle	329.5	245.7	415.0	321.7
Other	195.1	297.2	235.9	305.4
Suicide	320.8	388.2	359.2	391.7
Homicide	384.6	275.0	453.2	458.3
All other causes	119.2	169.9	105.3	177.4

Source: Robert D. Grove and Alice M. Hetzel, Vital Statistics Rates in the
United States, 1940-1960 (Washington, DC: Government Printing Office,
1968), Table 62; National Center for Health Statistics, "Advance Report of
Final Mortality Statistics, 1988," Monthly Vital Statistics Report 39:7
(Washington, DC: Government Printing Office, November 1990), Table 12.

[a]Deaths per 100,000 population in specified group.

while the age-adjusted death rates of black males for cancer of the urinary organs,
leukemia, and suicide were well below those of white males in both 1940 and
1988, the only cause in Table 5-11 for which the death rate dropped faster among
blacks than whites is pneumonia and influenza. Even in that case black males
are still more than half again as likely as whites to die from those diseases.
Females enjoyed more improvements and their racial gap in age-adjusted death
rates narrowed for genital cancer, pneumonia and influenza, ulcer, motor vehicle
accidents, and homicide. Nevertheless, the disadvantage of black women relative
to whites increased for cardiovascular diseases, most malignancies (including
breast cancer), diabetes, nonvehicular accidents, and suicide. This suggests that

Table 5-11
Black/White Age-Adjusted Mortality Ratio for Cause of Death, by Sex, 1940 and 1988

Cause of Death[a]	Male		Female	
	1940	1988	1940	1988
All causes	152.7	156.2	171.2	154.3
Tuberculosis	336.4	640.0	418.6	650.0
Malignant neoplasms	71.1	144.0	95.2	119.2
Digestive organs	76.0	159.6	71.8	145.1
Respiratory organs	54.8	143.8	67.6	99.2
Breast (female)	...	...	78.0	117.4
Genital organs	85.2	211.0	160.0	137.8
Urinary organs	57.1	85.7	70.7	120.7
Leukemia	57.4	92.1	47.1	92.1
Diabetes mellitus	72.8	206.3	98.2	263.1
Cardiovascular diseases	129.1	136.6	160.7	163.8
Pneumonia and influenza	219.1	155.6	222.6	125.2
Ulcer of stomach and duodenum	100.9	141.2	181.6	110.0
Appendicitis	105.5	400.0	121.6	200.0
Hernia and intestinal obstruction	148.4	210.0	149.4	160.0
Accidents	120.2	138.3	102.8	118.1
Motor vehicle	103.2	103.9	82.0	79.3
Other	131.6	184.1	108.8	179.2
Suicide	34.2	59.6	30.6	47.1
Homicide	1,142.0	755.8	969.2	453.6
All other causes	189.7	189.9	214.6	181.9

Source: Robert D. Grove and Alice M. Hetzel, Vital Statistics Rates in the United States, 1940-1960 (Washington, DC: Government Printing Office, 1968), Table 62; National Center for Health Statistics, "Advance Report of Final Mortality Statistics, 1988," Monthly Vital Statistics Report 39:7 (Washington, DC: Government Printing Office, 1990), Table 12.

[a]Deaths per 100,000 population in specified group.

as women move through young adulthood and middle age, the health gap between the races widens, reflected not only in differential death rates by cause, but also in differential rates of chronic hypertension, high cholesterol and sodium levels, alcohol abuse, anemia, and other health problems.[55] This widening gap in health status, which is closely associated with socioeconomic differences, helps explain why the death rates of very elderly blacks have risen for several causes, because aging becomes "an indicator of the length of exposure to life conditions that either undermine health (in the case of the disadvantaged) or promote it (in the case of the advantaged)."[56] Even so, the death rates of black females were below those of white females in both 1940 and 1988 for lung cancer, leukemia, motor vehicle accidents, and suicide. Nor has the mortality crossover in the oldest ages disappeared, and it even persists in some younger ages for certain causes of death.

 In summary, while the relative disadvantage of black women increased for some causes of death, the relative disadvantage of black men increased for virtually all of them. These and other comparisons show again that black males have the poorest potential to survive a host of diseases and environmental hazards.

NOTES

1. For a discussion of the methodology used in reporting cause of death, see National Center for Health Statistics, *Vital Statistics of the United States, 1987*, Vol. II, *Mortality*, Part A (Washington, DC: Government Printing Office, 1990), Sec. 7, pp. 8–11.

2. Alan D. Lopez, "Who Dies of What? A Comparative Analysis of Mortality Conditions in Developed Countries Around 1987," *World Health Statistical Quarterly* 43:2 (1990), pp. 105–114.

3. Robert D. Grove and Alice M. Hetzel, *Vital Statistics Rates in the United States: 1940–1960* (Washington, DC: Government Printing Office, 1968), pp. 79, 559–560.

4. Mary Catherine Bateson and Richard Goldsby, *Thinking A.I.D.S.* (Reading, MA: Addison-Wesley, 1988), p. 20.

5. Ibid., p. xv.

6. Charles F. Turner, Heather G. Miller, and Lincoln E. Moses, eds., *AIDS: Sexual Behavior and Intravenous Drug Use* (Washington, DC: National Academy Press, 1989), pp. 31–32.

7. U.S. Bureau of the Census, *Statistical Abstract of the United States: 1990* (Washington, DC: Government Printing Office, 1990), p. 116.

8. National Center for Health Statistics, *Health, United States, 1989* (Washington, DC: Government Printing Office, 1990), p. 151.

9. Turner et al., *AIDS*, p. 5.

10. Ibid., p. 2.

11. James W. Buehler, Owen J. Devine, Ruth L. Berkelman, and Frances M. Chevarley, "Impact of the Human Immunodeficiency Virus Epidemic on Mortality Trends in Young Men, United States," *American Journal of Public Health* 80:9 (September 1990), pp. 1080–1086.

12. See Matilda White Riley, Marcia G. Ory, and Diane Zablotsky, eds., *AIDS in an Aging Society: What We Need to Know* (New York: Springer, 1989).

13. Turner et al., *AIDS*, p. 3.

14. Ibid., pp. 3, 10. A comprehensive work on AIDS prevention is Ronald O. Valdiserri, *Preventing AIDS: The Design of Effective Programs* (New Brunswick, NJ: Rutgers University Press, 1989).

15. U.S. Bureau of the Census, "United States Population Estimates, by Age, Sex, Race, and Hispanic Origin: 1980 to 1988," *Current Population Reports*, Series P–25, No. 1045 (Washington, DC: Government Printing Office, 1990), pp. 19–20.

16. See Kathleen E. Ragland, Steve Selvin, and Deane W. Merrill, "The Onset of Decline in Ischemic Heart Disease Mortality in the United States," *American Journal of Epidemiology* 127:3 (March 1988), pp. 516–531.

17. For the supporting data, see NCHS, *Health, 1989*, pp. 169–170.

18. See U.S. Centers for Disease Control, "Smoking-Attributable Mortality and Potential Years of Life Lost—United States, 1984," *Morbidity and Mortality Weekly Report* 36:42 (October 1987), pp. 693–697.

19. For the data, see NCHS, *Health, 1989*, p. 165.

20. Hal B. Richardson, critique of Juha Pekkanen, Shai Linn, Gerardo Heiss, Chirayath M. Suchindran, Arthur Leon, Basil F. Rifkind, and Herman A. Tryoler, "Ten-Year Mortality from Cardiovascular Disease in Relation to Cholesterol Level Among Men With and Without Preexisting Cardiovascular Disease," *New England Journal of Medicine* 322:24 (June 14, 1990), pp. 1700–1717, in *New England Journal of Medicine* 324:1 (January 3, 1991), pp. 60–61. See also Keavan M. Anderson, William P. Castelli, and Daniel Levy, "Cholesterol and Mortality: Thirty Years of Follow-Up from the Framingham Study," *Journal of the American Medical Association* 257:16 (April 24, 1987), pp. 2176–2180.

21. Vaclav Smil, "Coronary Heart Disease, Diet, and Western Mortality," *Population and Development Review* 15:3 (September 1989), pp. 410–411. See also William P. Castelli, *The Fundamental Role of Serum Lipids in Coronary Heart Disease* (New York: Pfizer, 1984).

22. Smil, "Coronary Heart Disease," pp. 418–419.

23. For the data, see NCHS, *Health, 1989*, pp. 172–174.

24. For a more complete account of the disease, see Robert Berkow, ed., *The Merck Manual*, 14th ed. (Rahway, NJ: Merck Sharp & Dohme Research Laboratories, 1982), pp. 413–428.

25. For the data, see U.S. Bureau of the Census, "Money Income and Poverty Status in the United States: 1988," *Current Population Reports*, Series P–60, No. 166 (Washington, DC: Government Printing Office, 1989), Table 19.

26. For the data, see U.S. Bureau of the Census, "Measuring the Effect of Benefits and Taxes on Income and Poverty: 1989," *Current Population Reports*, Series P–60, No. 169-RD (Washington, DC: Government Printing Office, 1990), Table 2.

27. Bureau of the Census, "Population Estimates," Table D.

28. See Milton C. Weinstein, Pamela G. Coxson, Lawrence W. Williams, Theodore M. Pass, William B. Stason, and Lee Goldman, "Forecasting Coronary Heart Disease Incidence, Mortality, and Cost: The Coronary Heart Disease Model," *American Journal of Public Health* 77:11 (November 1987), pp. 417–426.

29. National Center for Health Statistics, Mary Grace Kovar, "Aging in the Eighties, People Living Alone—Two Years Later," *Advance Data from Vital and Health Statistics* 149 (Washington, DC: Government Printing Office, April 1988), p. 1. The earlier study in NCHS, Mary Grace Kovar, "Aging in the Eighties, Age 65 Years and Over and Living Alone, Contacts with Family, Friends, and Neighbors," *Advance Data from Vital and Health Statistics* 116 (Washington, DC: Government Printing Office, May 1986).

30. U.S. Bureau of the Census, Arlene F. Saluter, "Marital Status and Living Arrangements: March 1979," *Current Population Reports*, Series P–20, No. 349 (Washington, DC: Government Printing Office, 1980), Table 6; Bureau of the Census, Arlene F. Saluter, "Marital Status and Living Arrangements: March 1989," *Current Population Reports*, Series P–20, No. 445 (Washington, DC: Government Printing Office, 1990), Tables F, 7.

31. Eve Powell-Griner and Albert Woolbright, "Trends in Infant Deaths from Congenital Anomalies: Results from England and Wales, Scotland, Sweden and the United States," *International Journal of Epidemiology* 19:2 (June 1990), pp. 391–398.

32. Robert Adams, "Address to the American Heart Association Writers' Seminar." (Savannah, GA: January 16, 1991).

33. Walter C. Willett, Meir J. Stampfer, Graham A. Colditz, Bernard A. Rosner, and Frank E. Speizer, "Relation of Meat, Fat, and Fiber Intake to the Risk of Colon Cancer in a Prospective Study of Women," *New England Journal of Medicine* 323:4 (December 13, 1990), p. 1664.

34. Dwight T. Janerich, W. Douglas Thompson, Luis R. Varela, Peter Greenwald, Sherry Chorost, Cathy Tucci, Muhammed B. Zaman, Myron E. Melamed, Maureen Kiely, and Martin F. McKneally, "Lung Cancer and Exposure to Tobacco Smoke in the Household," *New England Journal of Medicine* 323:10 (September 6, 1990), p. 632.

35. NCHS, *Health, 1989*, Table 55.

36. Ibid., Table 53.

37. Jeff M. Hall, Ming K. Lee, Beth Newman, Jan E. Morrow, Lee A. Anderson, Bing Huey, and Mary-Claire King, "Linkage of Early-Onset Familial Breast Cancer to Chromosome 17q21," *Science* 250:4988 (December 21, 1990), p. 1684.

38. Brian Henderson, Annlia Paganini-Hill, and Ronald K. Ross, "Decreased Mortality in Users of Estrogen Replacement Therapy," *Archives of Internal Medicine* 151:1 (January 4, 1991), pp. 75–78.

39. See Claudia Wallis, "A Puzzling Plague," *Time* 137:2 (January 14, 1991), pp. 48–51.

40. Robert D. Retherford, *The Changing Sex Differential in Mortality* (Westport, CT: Greenwood Press, 1975), p. 14.

41. Phiroze Hansotia and Steven K. Broste, "The Effect of Epilepsy or Diabetes Mellitus on the Risk of Automobile Accidents," *New England Journal of Medicine* 324:1 (January 3, 1991), p. 25.

42. Berkow, *Merck Manual*, p. 1584.

43. For the social antecedents of suicide, see Emile Durkheim, *Suicide*, translated by John A. Spaulding and George Simpson (New York: Free Press, 1951); Jack P. Gibbs, ed., *Suicide* (New York: Harper & Row, 1969); Howard Gabennesch, "When Promises Fail: A Theory of Temporal Fluctuations in Suicide," *Social Forces* 67:1 (September 1988), pp. 129–145.

44. Rosemary Gartner, "The Victims of Homicide: A Temporal and Cross-National Comparison," *American Sociological Review* 55:1 (February 1990), pp. 100–102.

45. See also Dane Archer and Rosemary Gartner, *Violence and Crime in Cross-National Perspective* (New Haven, CT: Yale University Press, 1984); and Martin Daly and Margo Wilson, *Homicide* (Hawthorne, NY: Aldine de Gruyter, 1988).

46. Robert Nash Parker, "Poverty, Subculture of Violence, and Type of Homicide," *Social Forces* 67:4 (June 1989), pp. 988, 998–1000. See also Steven Messner, "Economic Discrimination and Societal Homicide Rates: Further Evidence on the Cost of Inequality," *American Sociological Review* 54:4 (August 1989), pp. 597–611.

47. Ibid., p. 597.

48. Rosemary Gartner and Robert Nash Parker, "Cross-National Evidence on Homicide and the Age Structure of the Population," *Social Forces* 69:2 (December 1990), pp. 360–361.

49. Gregory S. Kowalski and Don Duffield, "The Impact of the Rural Population Component on Homicide Rates in the United States: A County-Level Analysis," *Rural Sociology* 55:1 (Spring 1990), pp. 76–77. For an alternative analysis, see Kenneth P.

Wilkinson, "A Research Note on Homicide and Rurality," *Social Forces* 63:2 (December 1984), pp. 445–452.

50. Ibid., p. 450.

51. Kowalski and Duffield, "Rural Homicide," pp. 86–87.

52. Louis Henry, "Men's and Women's Mortality in the Past," *Population* 44:1 (September 1989), pp. 189–190.

53. Ibid., pp. 190–191.

54. Ibid., pp. 194–196.

55. Arline T. Geronimus and John Bound, "Black/White Differences in Women's Reproductive Health Status: Evidence from Vital Statistics," *Demography* 27:3 (August 1990), p. 463.

56. Ibid., p. 464. See also Arline T. Geronimus, "On Teenage Childbearing and Neonatal Mortality in the United States," *Population and Development Review* 13:2 (June 1987), pp. 245–279.

6 Life Expectancy

Chapter 1 discussed the concept and computation of life expectancy, emphasizing that the *current life table* represents a hypothetical situation that compares the risk of dying by age, sex, race, and geography.[1] The *generation life table* is also useful for certain purposes, since it is based on the actual mortality experience of a particular age cohort (e.g., all persons born in 1920), but requisite data are scarce and one must project life expectancy for cohort members still alive.[2] The current life table is more practical. Its underlying hypothesis assumes that if present age-specific mortality rates do not change, life expectancy expressed in the life table will be actual life expectancy at birth and other ages.[3] Chapter 1 also looked at the meaning of life table values necessary to compute expected number of years of life remaining at birth and other ages. We also projected ways life expectancy, and even life span, could change.

Several other aspects of life expectancy remain, including (1) differentials by age and sex; (2) differentials by age, sex, and race; (3) variations by state; (4) relationships between life expectancy and cause of death; (5) active life expectancy; (6) health goals and life expectancy; (7) long-term and short-term trends; and (8) some projections.

LIFE EXPECTANCY BY AGE AND SEX

For most purposes the abridged life table, based on life expectancy (or average remaining lifetime) in the ages 0–1, 1–5 . . . 85 and over, is acceptable. There are mortality data for the ages 85–100 and over by five-year age groups, but the numbers often are overstated and are small even if accurate.[4] Table 1–4 (Chapter 1) is an abridged life table (see especially column 7, or $\overset{o}{e}_x$). Table 6–1, however, includes only the values in column 7 of the abridged life table, so it gives life

Table 6-1

Life Expectancy at Birth and Percent Decrease at Successive Ages, All Races, by Sex, 1985-87

Age	Life Expectancy[a]			Percent Decrease		
	Both Sexes	Male	Female	Both Sexes	Male	Female
0-1	74.8	71.3	78.3	...	...	...
1-5	74.6	71.2	78.0	0.3	0.1	0.4
5-10	70.8	67.3	74.1	5.1	5.5	5.0
10-15	65.9	62.4	69.2	6.9	7.3	6.6
15-20	60.9	57.5	64.3	7.6	7.9	7.1
20-25	56.2	52.8	59.4	7.7	8.2	7.6
25-30	51.5	48.2	54.6	8.4	8.7	8.1
30-35	46.8	43.7	49.7	9.1	9.3	9.0
35-40	42.1	39.1	44.9	10.0	10.5	9.7
40-45	37.4	34.5	40.2	11.2	11.8	10.5
45-50	32.9	30.1	35.5	12.0	12.8	11.7
50-55	28.5	25.8	31.0	13.4	14.3	12.7
55-60	24.3	21.8	26.6	14.7	15.5	14.2
60-65	20.4	18.0	22.5	16.1	17.4	15.4
65-70	16.8	14.7	18.6	17.6	18.3	17.3
70-75	13.6	11.7	15.0	19.0	20.4	19.4
75-80	10.7	9.1	11.7	21.3	22.2	22.0
80-85	8.1	6.9	8.8	24.3	24.2	24.8
85 and over	6.0	5.2	6.4	25.9	24.6	27.3

Source: National Center for Health Statistics, Vital Statistics of the United States, 1985, 1986, 1987, Vol. II, Mortality, Part A (Washington, DC: Government Printing Office, 1988-90), Sec. 6, Table 1.

[a]Years of life remaining at beginning of age interval.

expectancy by the specific age groups in Table 1-4 and the percentage decreases in life expectancy from one age group to the next. Life expectancy is expressed as the years of life remaining at the beginning of each age interval, which is why the intervals appear to overlap (e.g., 0-1, 1-5, 5-10). Furthermore, Table 6-1 is for males and females of all races collectively and for 1985-87 so as to minimize the effect of annual fluctuations.

For most groups in the United States, life expectancy falls in each succeeding age category, but the pace of reduction rises steadily with age, until the decrement in the ages 85 and over is five times that in the ages 5-10. The drop in life expectancy between ages 0-1 and 1-5 is small, partly because there is only one year between the beginning of the first interval and the beginning of the second, but partly because death rates are much higher among infants than children 1-5 and life expectancy in both groups is similar. Life expectancy then falls by larger percentages in succeeding age groups, until it is 25 percent less for persons aged 85 than for those aged 80. This reflects rising death rates with age, and while overall life expectancy increased dramatically in the twentieth century, the increase was due more to the growing proportion of infant lives saved than to prolonging life among the elderly, though each change played a part in the overall increase. For example, while the death rate among infants fell 80 percent between 1940 and 1987, the rate among persons aged 65 dropped only 44 percent. Both are significant reductions, but the greater decrement of change in the younger ages belies the common assumption that life expectancy has risen essentially because death rates of older people have dropped. On the other hand, the longer one survives, the more one's years already lived plus life expectancy increase over the life expectancy of a newborn. In 1985–87, for example, a newborn had an average life expectancy of 74.8 years, but a person who lived to age 20 could expect an average of 56.2 more years, for a total of 76.2. At age 65 the years lived plus average life expectancy totaled 81.8 years; at age 85 it totaled 91.0 years.

Many of these patterns apply to both sexes, although at most ages life expectancy declines faster among males than females. The few exceptions include the time between infancy and early childhood, when life expectancy decreases slightly more for females than males. At age 80 and over life expectancy also declines somewhat faster for females than males, and while 2.6 years of life expectancy separate the sexes at age 75, by age 85 the difference is only 1.2 years (see Table 6–2). This is expected, given the more rapid increase, or slower decrease, in death rates for certain kinds of cancer among women than men, discussed in Chapter 5, along with slower decreases among white women in death rates for certain kinds of heart disease. Postmenopausal women lose much of the survival advantage provided by estrogen, and while estrogen replacement helps preserve some of that advantage, it does not prevent convergence in the life expectancy of men and women as they age. Thus, even though life expectancy generally falls somewhat faster with age among males than females, elderly women have a much smaller advantage over elderly men than girl infants have over boy infants. At birth females can expect to live 7.0 years longer than males, but the gap narrows steadily to only 1.2 years at age 85. Nonetheless, trends since the turn of the century do show a growing advantage of females over males in life expectancy at birth.[5]

LIFE EXPECTANCY BY AGE, SEX, AND RACE

Most patterns just discussed apply to blacks and whites, although blacks have significantly lower life expectancy than whites at all ages until about 80, after which convergence and crossover are reported (see Tables 6–2 and 6–3). Furthermore, life expectancy falls faster among black than white males until ages 45–50, after which the drop is faster among whites, particularly in the oldest ages. The same situation applies to females until age 45, after which the drop is faster for whites. These reversals generally reflect movement toward crossover in the oldest ages, although it has earlier antecedents for certain causes of death.

Table 6–2
Difference Between Males and Females and Blacks and Whites in Years of Life Expectancy at Birth, by Age, 1985–87

Age	Male/Female Difference			Black/White Difference		
	All Races	White	Black	Both Sexes	Male	Female
0-1	7.0	6.8	8.3	6.0	6.8	5.3
1-5	6.8	6.7	8.2	5.4	6.1	4.6
5-10	6.8	6.7	8.2	5.4	6.1	4.6
10-15	6.8	6.6	8.2	5.4	6.1	4.5
15-20	6.8	6.6	8.2	5.3	6.1	4.5
20-25	6.6	6.5	8.0	5.3	6.1	4.6
25-30	6.4	6.2	7.6	5.2	5.9	4.5
30-35	6.0	5.9	7.3	5.0	5.7	4.3
35-40	5.8	5.8	6.8	4.7	5.2	4.2
40-45	5.7	5.6	6.3	4.2	4.6	3.9
45-50	5.4	5.4	5.8	3.8	4.0	3.6
50-55	5.2	5.2	5.3	3.3	3.3	3.2
55-60	4.8	4.9	4.7	2.7	2.6	2.8
60-65	4.5	4.4	4.2	2.2	2.1	2.3
65-70	3.9	3.9	3.6	1.5	1.4	1.7
70-75	3.3	3.4	3.1	1.1	0.9	1.2
75-80	2.6	2.7	2.4	0.6	0.4	0.7
80-85	1.9	1.9	1.8	0.2	0.1	0.2
85 and over	1.2	1.3	1.2	+0.4[a]	+0.5[a]	+0.4[a]

Source: National Center for Health Statistics, Vital Statistics of the United States, 1985, 1986, 1987, Vol. II, Mortality, Part A (Washington, DC: Government Printing Office, 1988-90), Sec. 6, Table 1.

[a]Reported life expectancy of blacks higher than that of whites.

Due to the reversals, black men aged 85 have 0.5 year more life expectancy than white men, and black women have 0.4 year more than white women. Thus, while the average black male newborn can expect to live 6.9 years less than the average white newborn, by age 85 that significant disadvantage has given way to a small advantage. The average black female newborn has 5.3 fewer years of life expectancy than her white counterpart, but by age 85 that liability has become a small asset. As noted earlier, the crossover effect is a complex phenomenon that calls for caution because at least part is due to overstatement of ages in the advanced years—a situation Ansley Coale and Ellen Kisker found virtually universal even in countries with excellent censuses and registrations.[6]

Table 6–3 also shows the great risks of being a black newborn, male or female, because life expectancy among infants is actually lower than in the ages 1–5. Thus, a black male newborn has 0.4 year less life expectancy than a black male who has reached age 1; among females the difference is 0.3 year. No comparable situation exists for whites because their hazards in the first year are less likely to be fatal, so white life expectancy is highest at birth and then decreases steadily.

Table 6–3
Life Expectancy at Specified Ages, by Race and Sex, 1985–87[a]

Age	White			Black		
	Both Sexes	Male	Female	Both Sexes	Male	Female
0–1	75.4	72.0	78.8	69.4	65.2	73.5
1–5	75.1	71.7	78.4	69.7	65.6	73.8
5–10	71.3	67.9	74.6	65.9	61.8	70.0
10–15	66.4	63.0	69.6	61.0	56.9	65.1
15–20	61.4	58.1	64.7	56.1	52.0	60.2
20–25	56.7	53.4	59.9	51.4	47.3	55.3
25–30	52.0	48.8	55.0	46.8	42.9	50.5
30–35	47.2	44.2	50.1	42.2	38.5	45.8
35–40	42.5	39.5	45.3	37.8	34.3	41.1
40–45	37.8	34.9	40.5	33.6	30.3	36.6
45–50	33.2	30.4	35.8	29.4	26.4	32.2
50–55	28.8	26.0	31.2	25.5	22.7	28.0
55–60	24.5	21.9	26.8	21.8	19.3	24.0
60–65	20.6	18.2	22.6	18.4	16.1	20.3
65–70	16.9	14.8	18.7	15.4	13.4	17.0
70–75	13.6	11.7	15.1	12.5	10.8	13.9
75–80	10.7	9.1	11.8	10.1	8.7	11.1
80–85	8.1	6.9	8.8	7.9	6.8	8.6
85 and over	6.0	5.1	6.4	6.4	5.6	6.8

Source: National Center for Health Statistics, Vital Statistics of the United States, 1985, 1986, 1987, Vol. II, Mortality, Part A (Washington, DC: Government Printing Office, 1988–90), Sec. 6, Table 1.

[a]Life expectancy is years of life remaining at beginning of age interval.

The male/female differential in life expectancy is consistently greater for blacks than whites at all ages up to 55. This reflects the large mortality disadvantage among black male children, adolescents, young adults, and middle-aged men that has many causes and manifestations. Thus, at birth white females can expect to live 6.8 years longer than white males, but black females can expect to live 8.3 more years than black males (see Table 6–2). The difference persists, though it shrinks, until age 55, after which the sex gap in life expectancy is less for blacks than whites; then both races experience increasing homogenization in life expectancy with age. In fact, race and sex differences in life expectancy are significantly smaller in the oldest ages than nearer the lower end of the age scale. Thus, the classic advantages in mortality and life expectancy—females over males and whites over blacks—diminish and even disappear in the elderly population as various homogenizing factors operate, although the relationship between life expectancy and death rates at given ages is complex and needs to account for specific causes.[7] In addition, given the quality of mortality data in the oldest ages, the homogenizing effect, too, should be treated cautiously.[8]

Some actual homogenization does occur, however, partly because of biological factors: male and female death rates converge for many causes due to women's menopause; selection produces a somewhat more durable black than white elderly population. Environmental factors also cause more similar risks for elderly blacks and whites of both sexes. Few persons in their 70s and 80s, for example, are in the labor force and the tendency for men and blacks to have more hazardous jobs than women and whites has little effect on mortality, though it may have some residual influence. More similar behavior also reduces the large differentials in death rates for accidents, suicide, and homicide. Finally elderly persons, regardless of race and sex, become increasingly alike in their susceptibility to heart disease, cancer, stroke, and other causes. On balance, therefore, many sharp differences in life expectancy among younger populations decrease substantially among older groups.

Survivors at Specified Ages

Another measure of differentials and convergence is the percentage of persons born alive who are still living at successive age intervals. This is derived from column 3 (l_x) of the life table, which gives the number of persons living at the beginning of each age interval out of each 100,000 born alive, though columns l_x and n^{d_x} could be used to answer other questions about the proportion of persons living or dying.[9] Table 6–4 gives the percentage of survivors at the beginning of each age interval in 1985–87 for blacks and whites, males and females; it shows how heavily deaths are concentrated in the older ages and how one's survival prospects vary by race and sex. For example, over 90 percent of the original cohort of white females was still alive at age 60; but white males and black females have 90 percent survivorship only until age 50, black males only until age 35. Furthermore, the percentage of survivors is smaller among white males than white females at every age and the gap widens with age, until at 85 and over the proportion of females still living is twice that among males. Similar male/female differences in survivorship appear among blacks, but at most ages the sex gap is even wider than among whites.

Since survivorship rates are lower among blacks than whites and among males than females, why does the racial mortality crossover occur in the oldest years and why is there a tendency toward homogenization? Much of the answer lies in the different survival potential of white males and black females. That is, the percentage of survivors in each age interval after 70 is significantly higher among black females than among white males. Therefore, even though successive percentages of survivors are greater among white males than black males and among white females than black females, the greater durability of elderly black women than elderly white men is responsible for much of the mortality crossover.[10] This is true not only for all causes of death collectively, but also for each of the major ones individually, except for diabetes mellitus, hypertensive heart disease in the ages 80 and over, stroke in the ages 85 and over, and, of course, breast cancer.[11]

Table 6–4
Percent of Persons Born Alive and Still Living at the Beginning of Each Age Interval, by Race and Sex, 1985–87

Age	White		Black	
	Male	Female	Male	Female
0–1	100.00	100.00	100.00	100.00
1–5	98.99	99.22	98.01	98.38
5–10	98.79	99.06	97.67	98.10
10–15	98.66	98.97	97.48	97.95
15–20	98.49	98.88	97.26	97.84
20–25	97.92	98.64	96.61	97.61
25–30	97.14	98.39	95.42	97.22
30–35	96.39	98.13	93.92	96.66
35–40	95.55	97.80	91.86	95.80
40–45	94.54	97.33	89.18	94.66
45–50	93.17	96.59	85.72	92.99
50–55	91.09	95.38	81.36	90.61
55–60	87.73	93.41	75.38	87.05
60–65	82.48	90.39	67.93	82.08
65–70	74.80	85.73	58.20	74.98
70–75	64.60	79.10	47.40	66.46
75–80	51.21	69.66	35.32	55.99
80–85	35.89	56.87	23.63	43.14
85 and over	20.63	40.16	12.70	28.14

Source: National Center for Health Statistics, Vital Statistics of the United States, 1985, 1986, 1987, Vol. II, Mortality, Part A (Washington, DC: Government Printing Office, 1988–90), Sec. 6, Table 1.

Differential Life Expectancy and Widowhood

Greater life expectancy and higher rates of survivorship among females cause widows to outnumber widowers at every age, although widowers also are considerably more likely than widows to remarry. In the older ages the difference between the number of widows and widowers is substantial, as are the socioeconomic implications for this large number of women who have outlived their spouses. In 1989 about three-quarters of all men 65 and over were married and living with a spouse, but that was true for only two-fifths of the women, mostly because of widowhood and different remarriage rates, although a somewhat larger percentage of women than men had never married (see Table 6–5). Nearly half the women 65 and over were widows, whereas only 14 percent of the men were widowers. Moreover, these sex differences increase with age, until 82 percent of women 85 and over must contend with widowhood, nearly all without the prospect of remarriage. Even in those ages 48 percent of men are still married, 42 percent widowed.[12] The wide sex gap in the incidence of widowhood among the elderly applies to blacks and whites at all ages 65 and over. Black women

Table 6–5
Percent of Older Persons Classified as Widowed, by Race and Sex, 1989

Race and Sex	65+ Years	65–74 Years	75–84 Years	85+ Years
All races				
Male	14.0	8.9	19.7	42.1
Female	48.7	36.6	61.5	82.3
White				
Male	13.2	8.1	19.2	41.8
Female	48.1	35.5	61.0	82.7
Black				
Male	21.4	17.6	24.7	43.5
Female	55.3	47.1	66.9	76.3

Source: U.S. Bureau of the Census, Arlene F. Saluter, "Marital Status and Living Arrangements: March 1989," Current Population Reports, Series P-23, No. 445 (Washington, DC: Government Printing Office, 1990), Table 1.

are more likely than white women to be widows in the ages 65–84, however, but somewhat less likely in the ages 85 and over.

Despite these results of sex differentials in life expectancy, long-term changes in fertility and mortality allow the average young couple who marry today and produce children to expect to live together 14 years longer than their parents did after the last child has left home, barring divorce or separation.[13] Those 14 years represent about a third of the 44 years that persons with continuous first marriages can expect to spend together.[14] Therefore, while widowhood is tragic for individuals, its patterns represent substantial improvements in life expectancy that allow a couple to spend many older years together free of child care. At the same time, the numbers of widowed older men and women are greatly out of balance, and men have a much better statistical chance for remarriage. In 1989, for example, among widowed persons 65 and over, there were only 20 men per 100 women, and among those in all unmarried categories the ratio was just 28. Moreover, widowers average three years older than widows, so differential life expectancy means that most elderly widows who do remarry can expect to become widows again.

Widows and widowers generally marry each other rather than divorced or single persons, though they have to choose within a fairly narrow age range because norms still keep most women from marrying much younger men. Those constraints further limit a widow's choices and discourage many from marrying no matter what their statistical prospects. Furthermore, the older a woman is when she becomes a widow, the smaller her chances of remarriage, not only because of the sex ratio, but because many social factors conspire against her.[15] Younger widows are especially likely to remarry if they have worked for wages, are adept at making personal adjustments, and had deeply satisfying emotional relationships with their deceased husbands.[16] Many factors reduce the number of elderly people who would remarry even if the sex ratio were closer to 100, so the 7.1 million "surplus" unmarried women 65 and over are not an eager throng who would all choose new husbands if only there were enough to go around.[17] In any case, for many elderly people sex variations in life expectancy contribute to loneliness, economic deprivation, dilemmas over remarriage, and other social and psychological problems.

GEOGRAPHIC DISTRIBUTION OF LIFE EXPECTANCY

Life expectancy varies considerably by state, basically for reasons discussed under differential mortality by state. They include proportional representation of various racial and ethnic groups with high or low death rates, quality and availability of local health and medical care, income and poverty levels, and other forces that influence mortality and, therefore, life expectancy. Table 6–6 shows life expectancy at birth for the total population and for black and white males and females in each state and the District of Columbia; the states are ranked according to longevity in their total populations. The columns for blacks do not include data for every state because some have too few blacks to enable accurate comparisons. In fact, small numbers also make it difficult to compute life expectancy in smaller units than the state, even for all races combined.[18] In addition, the data are for 1979–81 because of the unavoidable time lag in gathering data, reporting results, and making computations. After the results of a decennial census are available, the NCHS computes life expectancy for states, using death rates for the census year and one year on each side of it (e.g., 1979–81). This procedure uses the most accurate figures available because populations in a census year are enumerated, whereas those in intercensal years are estimated and grow less accurate the more remote they are from a census year and the more they are subdivided by race, sex, and other characteristics. At this writing the figures on life expectancy for 1979–81 are the most accurate available by race and sex in the states, so they are used to show state-to-state variations, though obviously not changes since then.

Hawaii ranks highest in average life expectancy at birth, the District of Columbia lowest—relative positions the two occupy on many measures of mortality and well-being. Hawaii and the District of Columbia are separated by 7.8 years of life expectancy, largely because of socioeconomic variations among their constituent races. In Hawaii persons of Japanese ancestry are nearly a quarter of the total population, and their death rates are exceptionally low, their life expectancy unusually high. The same is true for whites, who are a third of the population, and for persons with Chinese and certain other ethnic backgrounds. In the District of Columbia blacks are nearly 70 percent of the total, and their death rate is comparatively high, their life expectancy relatively low, so the District of Columbia is the only unit in Table 6–6 with life expectancy of less

Table 6–6
Years of Life Expectancy at Birth, by Race, Sex, and State, 1979–81

State	Rank	All Races	White Male	White Female	Black Male	Black Female
United States	...	73.88	70.82	78.22	64.10	72.88
Hawaii	1	77.02	73.04	79.81	a	a
Minnesota	2	76.15	72.63	79.90		
Iowa	3	75.81	72.09	79.64	a	a
Utah	4	75.76	72.42	79.22	a	a
North Dakota	5	75.71	72.45	79.95	a	a
Nebraska	6	75.49	71.97	79.53	a	a
Wisconsin	7	75.35	72.05	79.05	66.98	74.09
Kansas	8	75.31	71.85	79.26	66.17	73.24
Colorado	9	75.30	71.84	78.89	67.41	74.66
Idaho	10	75.19	71.58	79.19	a	a
Washington	11	75.13	71.86	78,64	a	a
Connecticut	12	75.12	71.90	78.86	65.80	74.62
Massachusetts	13	75.01	71.38	78.54	67.53	75.73
Oregon	14	74.99	71.41	78.79	a	a
New Hampshire	15	74.98	71.39	78.38	a	a
South Dakota	16	74.97	72.07	80.07	a	a
Vermont	17	74.79	71.03	78.47	a	a
Rhode Island	18	74.76	71.06	78.45	a	a
Maine	19	74.59	70.77	78.39	a	a
California	20	74.57	71.18	78.12	65.47	73.74
Arizona	21	74.30	71.08	78.66	a	a
New Mexico	22	74.01	70.46	78.63	a	a
Florida	23[b]	74.00	71.10	78.86	63.05	71.79
New Jersey	23[b]	74.00	71.25	77.99	64.53	73.02
Montana	25	73.93	71.00	78.19	a	a
Wyoming	26	73.85	70.15	78.39	a	a
Indiana	27[b]	73.84	70.57	77.82	64.71	72.87
Missouri	27[b]	73.84	70.64	78.29	63.14	72.65
Arkansas	29	73.72	70.46	78.59	65.00	73.77
New York	30	73.70	70.90	77.80	64.14	73.28
Michagan	31[b]	73.67	70.94	77.99	63.87	72.58
Oklahoma	31[b]	73.67	69.90	78.07	64.71	73.22
Texas	33	73.67	70.30	78.22	64.44	73.42
Pennsylvania	34	73.58	70.52	77.64	63.27	72.35
Ohio	35	73.49	70.42	77.53	64.56	72.75
Virginia	36	73.43	70.54	78.28	65.08	72.99
Illinois	37	73.37	70.57	77.96	63.02	72.09
Maryland	38	73.32	70.86	77.73	65.13	73.25
Tennessee	39	73.30	69.99	78.31	64.07	72.96
Delaware	40	73.21	70.53	77.59	64.35	72.53

Table 6–6 (continued)

State	Rank	All Races	White		Black	
			Male	Female	Male	Female
Kentucky	41	73.06	69.46	77.46	64.31	72.38
North Carolina	42	72.96	70.02	78.53	63.33	73.32
West Virginia	43	72.84	68.99	77.09	63.66	71.94
Nevada	44	72.64	69.52	76.73	a	a
Alabama	45	72.53	69.67	78.15	63.54	72.89
Alaska	46	72.24	69.99	77.93	a	a
Georgia	47	72.22	69.56	78.01	63.18	71.88
Mississippi	48	71.98	69.26	78.09	64.09	73.32
South Carolina	49	71.85	69.40	77.81	62.73	72.31
Louisiana	50	71.74	69.20	77.42	63.29	72.27
District of Columbia	51	69.20	71.24	77.88	61.88	72.01

Source: National Center for Health Statistics, U.S. Decennial Life Tables for 1979–81, Vol. I, No. 4 (Washington, DC: Government Printing Office, 1988), Table C.

aInsufficient data.

bStates tied for rank.

than 70 years. In fact, the gap in longevity between blacks and whites in the District of Columbia is quite wide, and while life expectancy for blacks is lower than in any state, life expectancy for whites is higher than in 32 states. Therefore, whites in the District of Columbia can expect to live an average of 7.9 years longer than blacks.

Most southern states rank relatively low in life expectancy, although Nevada and Alaska are not far from the bottom. Others below the national average include several where heavy industry long dominated their economies, where recessions and unemployment have hit hard, and where large metropolitan centers have high percentages of poor persons, especially blacks. Several states in the Midwest and the West have the highest average longevity, and all six New England states rank above the national average. In all cases, the socioeconomic aspects of racial and ethnic composition affect longevity, just as they do state-to-state variations in age-adjusted death rates, discussed earlier.

Differentials by Sex

The life expectancy gap between males and females is substantial in all states. It ranges from 6.3 years in Hawaii to 9.2 years in the District of Columbia. Those units are also at the extremes of overall life expectancy and illustrate the tendency for the largest sex gap to appear where life expectancy is lowest, the smallest gap to show up where life expectancy is highest. For example, the ten states with highest life expectancy have an average sex gap of 7.2 years, whereas the ten with lowest life expectancy (including the District of Columbia) have a sex gap of 8.4 years. Thus, while females have a large longevity advantage in every state, males are especially short lived where many persons of both sexes lack good medical care, disease-prevention knowledge and programs, protection from accidents and violence, and other death controls.

These patterns by sex apply to blacks and whites. The male/female gap in life expectancy among whites ranges from 6.6 years in the District of Columbia to 8.8 in Mississippi. At first glance this ranking for the District of Columbia seems inconsistent with the association between a large sex gap in longevity and low overall longevity. But the District of Columbia's overall low life expectancy is due to the dismal showing of the black population and the relatively low life expectancy of its white females, despite the relatively high life expectancy among white males. In fact, they have higher life expectancy in the District of Columbia than in 33 states, whereas white women in the District of Columbia have higher life expectancy than in 11 states. Therefore, the relatively narrow sex gap among whites is due to a smaller mortality differential than in much of the rest of the nation. Apart from this situation, the smallest sex gaps in life expectancy among whites do appear in states with the longest life expectancy for their race, the largest gaps where life expectancy is shortest. Thus, the ten states with high life expectancy for whites have a sex gap in longevity of 7.3 years, while the ten states with low life expectancy have a sex difference of 8.2 years.

The situation is similar for blacks. Their highest life expectancy is in Massachusetts—one of only four states in which black longevity at birth exceeds 70 years. In Massachusetts the sex difference is 8.2 years, while it is 10.1 years in the District of Columbia, which has the nation's lowest life expectancy. The ten states with the longest life expectancy for blacks have a sex difference of 8.1 years, the ten with the shortest life expectancy have a gap of 9.1 years. These gaps, however, are larger than in the white population for comparable categories of states, and they reflect the serious disadvantage of black males, whose life expectancy is by far the lowest of the four race and sex groups.

Differentials by Race

The relative disadvantage of blacks, especially males, is further underscored by black/white longevity differentials for the states, although the comparison is confined to the District of Columbia and 30 states because of relatively small numbers of blacks in the others. Table 6–7 shows the number of years by which the life expectancy of whites, separately by sex, exceeds that of blacks in each state. It also shows substantial variations, from 7.9 years in the District of Columbia to only 3.4 years in Massachusetts, where various programs and other means of reducing death rates are more readily available to the several races. In every state except Kansas the longevity gap between blacks and whites is

Table 6-7
Years by Which Life Expectancy at Birth Among Whites Exceeds That Among
Blacks, Selected States, by Sex, 1979-81

State	Both Sexes	Male	Female
United States	6.01	6.72	5.34
District of Columbia	7.87	9.36	5.87
Florida	7.56	8.05	7.07
Illinois	6.66	7.55	5.87
Missouri	6.52	7.50	5.64
Michigan	6.27	7.07	5.41
Pennsylvania	6.24	7.25	5.29
Georgia	6.14	6.38	6.13
South Carolina	6.02	6.67	5.50
North Carolina	5.96	6.69	5.21
Kansas	5.89	5.68	6.02
New Jersey	5.82	6.72	4.97
Delaware	5.73	6.18	5.06
Alabama	5.55	6.13	5.26
Tennessee	5.53	5.92	5.35
New York	5.47	6.76	4.52
Virginia	5.46	5.46	5.29
Indiana	5.44	5.86	4.95
Louisiana	5.41	5.91	5.15
Ohio	5.34	5.86	4.78
Texas	5.34	5.86	4.80
Maryland	5.19	5.73	4.48
Connecticut	5.14	6.10	4.24
California	5.13	5.71	4.38
Kentucky	5.07	5.15	5.08
West Virginia	5.07	5.33	5.15
Wisconsin	5.00	5.07	4.96
Oklahoma	4.97	5.19	4.85
Arkansas	4.95	5.46	4.82
Mississippi	4.80	5.17	4.77
Colorado	4.36	4.43	4.23
Massachusetts	3.40	3.85	2.81

Source: National Center for Health Statistics, U.S. Decennial Life Tables
for 1979-81, Vol. I, No. 4 (Washington, DC: Government Printing Office,
1988), Table C.

greater for males than females. Among males the greatest racial differential is
in the nation's capital, the least in Massachusetts, but among females the largest
gap occurs in Florida. It reflects Florida's wide socioeconomic difference between
the elderly white female population, a large share of which is relatively affluent
in-migrants, and the elderly black female group, a larger share of which is
indigenous to Florida and still enmeshed in poverty, both rural and urban. Much
the same applies to black males in Florida, for their average life expectancy at

birth is 8.1 years less than that of white males. At age 65 white males in Florida have a life expectancy of 15.5 years, black males 13.1 years, for a difference of 2.4 years, contrasted with 1.0 year in the nation as a whole. Florida's white females aged 65 and over have a life expectancy of 19.5 years, blacks 17.0 years, for a difference of 2.5 years, compared with 1.4 years in the nation as a whole.[19] In many ways Florida is a very different place socioeconomically for blacks and whites, and while the same is true in other parts of the nation (e.g., Illinois, Missouri, Michigan), the contrast is not as sharp.

Despite the narrowing black/white difference over time, in 1979–81 the state variations coalesced into life expectancy for blacks at the national level that was about the same as that for whites in the 1950s.[20] By 1988 life expectancy among black males (64.9 years) was roughly equal to that of white males in 1946; that among black females (73.4 years) was about the same as life expectancy for white females in 1954. Furthermore, life expectancy for black males in 1988 was no better than that for white females in 1935, although longevity among black females first exceeded that among white males after 1955.[21] These data reflect serious inequities between black males and the rest of the population, not just in life expectancy, but in the quality of life that affects life expectancy. Black males fare so poorly that their average life expectancy at birth is below that in the total populations of 85 other countries and less than a year above that for the world as a whole. The deficiencies are reflected in comparatively high age-adjusted death rates among black males for practically all leading causes of death, but especially cardiovascular diseases, cancer, and homicide.[22] Furthermore, while the racial differential in longevity did decline, the pace of decline has slowed, thus allowing whites to maintain their long-time lead over blacks.[23] In fact, NCHS data show that the life expectancy of blacks has even gone down since 1984, while that of whites has continued to rise. In that year, newborn black males could expect to live an average of 65.6 years, but life expectancy declined slightly in each succeeding year, to 64.9 years in 1988. The reduction for black females was from 73.7 years to 73.4 years.[24] Except for a few causes of death, the mortality rates among black males and females are higher than those among white males and females throughout the country, thus causing blacks' consistently lower average life expectancy in every state shown in Table 6–7.

LIFE EXPECTANCY AND CAUSE OF DEATH

Eliminating Specific Causes of Death

An intriguing question in the study of longevity is how much it would increase if certain causes of death were eliminated, if, for example, we found complete cures for cancer or heart disease or could eradicate violence from American culture. Studies have charted life expectancy by hypothetically eliminating various causes, although it is obviously absurd to eliminate all of them statistically, since presumably people would never die.[25] In performing this exercise one has to account for age-specific death rates by cause, eliminate from the life table the rate for each cause, and compute life expectancy on that basis. The NCHS makes these calculations for the three years centered on a decennial census, so for 1979–81 it is possible to ascertain life expectancy at various ages if 14 major causes (and some subcategories) are eliminated. Those causes are listed in Table 6–8. These manipulations show prospective longevity for the age groups 0–1, 1–5 . . . 100 and over, but in this section we consider only life expectancy at birth, separately by race and sex.[26]

Table 6–8 shows the number of years added to life expectancy at birth when specific causes are eliminated. Naturally, populations with relatively high mortality rates for specific causes experience the largest gains in life expectancy from hypothetically eliminating those causes; this pinpoints areas of greatest need.[27] It should be noted too, however, that the age pattern of a disease also influences the gain in life expectancy from elimination of the disease. Since heart disease is by far the major cause of death among both sexes and all races, its elimination adds most to life expectancy—between 5.2 and 6.2 years, depending on race and sex. If a particular population has high death rates for several causes, however, eliminating any one of them produces a smaller gain in life expectancy than if a few causes claim most lives and one of those causes is eliminated. In high-mortality groups there are simply so many other threats to life that eliminating one cause may have only a modest impact on life expectancy.[28] This is why the potential gain from eliminating heart disease is smallest for black males (5.2 years), even though they have the highest age-adjusted death rate for heart disease. Their death rates are also relatively high for so many other causes that their life expectancy is short compared with black females and whites of both sexes, even with the savings from eliminating heart disease. In 1988, for example, heart disease accounted for 27 percent of deaths among black males, but 33 percent among black females. Consequently, females gained more years of life expectancy from eliminating heart disease, even though the death rate for it is higher among males than females.

All groups also gain significantly by eliminating cancer, and in this case black males gain most because their death rate for that cause is much higher than those in any other race and sex group. So, four factors affect the number of years added to life expectancy by eliminating a specific cause of death: (1) how high the death rate is for that cause, (2) how large a percentage of all deaths result from that cause, (3) whether the population experiencing the savings has high or low overall mortality, and (4) the average age of death for a cause. If overall mortality is high, even if persons are saved from one cause, they are likely to succumb relatively soon to another. These factors also explain why the gain in life expectancy by eliminating homicide is several times greater for black males than for any other group because their losses to that cause are exceptionally high. The same forces produce a greater gain in life expectancy among women than men when cerebrovascular diseases are eliminated.

Table 6–8

Years Added to Life Expectancy at Birth by Eliminating Selected Causes of Death, by Race and Sex, 1979–81

Cause	All Races	White		Black	
		Male	Female	Male	Female
Diseases of heart	5.79	5.83	5.31	5.18	6.15
Ischemic	3.80	4.15	3.35	2.69	3.03
Hypertensive	0.12	0.08	0.11	0.28	0.38
Malignant neoplasms	3.00	2.86	2.98	3.22	2.97
Respiratory organs	0.75	0.95	0.46	1.04	0.40
Digestive organs	0.69	0.63	0.67	0.77	0.76
Genital organs	0.28	0.20	0.34	0.32	0.41
Breast	0.26	0.00	0.59	0.00	0.51
Leukemia	0.12	0.12	0.09	0.09	0.09
Urinary organs	0.10	0.12	0.08	0.08	0.07
Accidents	1.14	1.52	0.66	1.57	0.65
Motor vehicle	0.62	0.87	0.37	0.65	0.23
Other	0.50	0.63	0.28	0.90	0.42
Cerebrovascular diseases	0.91	0.63	1.07	0.97	1.50
Certain conditions of perinatal period	0.47	0.43	0.37	0.84	0.79
Chronic obstructive pulmonary diseases	0.32	0.39	0.23	0.25	0.13
Pneumonia and influenza	0.28	0.24	0.27	0.38	0.30
Suicide	0.28	0.42	0.17	0.22	0.06
Homicide	0.28	0.25	0.09	1.44	0.36
Chronic liver disease and cirrhosis	0.26	0.28	0.18	0.46	0.30
Congenital anomalies	0.25	0.25	0.25	0.23	0.24
Diabetes mellitus	0.20	0.15	0.23	0.22	0.47
Infectious and parasitic diseases	0.13	0.10	0.11	0.24	0.25
Septicemia	0.06	0.04	0.05	0.10	0.13
Tuberculosis	0.01	0.01	0.06	0.06	0.03
Atherosclerosis	0.12	0.08	0.15	0.07	0.14

Source: National Center for Health Statistics, Lester R. Curtin and Robert J. Armstrong, "United States Life Tables Eliminating Certain Causes of Death," U.S. Decennial Life Tables for 1979–81, Vol. I, No. 2 (Washington, DC: Government Printing Office, 1988), Table D.

Eliminating each of many other causes adds little to life expectancy. Except for heart disease, cancer, and stroke among blacks and whites of both sexes, accidents among males of both races, and homicide among black males, eliminating other specific causes saves less than one year of life; in most cases it saves less than a half year. In fact, any future increases in life expectancy will have to come largely from reductions in the death rates for heart disease and cancer because they cause almost two-thirds of all deaths.[29] Even stroke runs a distant third in this respect.

Gains in life expectancy by eliminating particular causes also vary according to the average age at death from those causes; eliminating any that take their largest toll early in life (e.g., congenital anomalies) produces larger gains than eliminating causes that usually occur much later (e.g., heart disease).[30] For this reason, at age 35 even eliminating heart disease adds only about three years to the average American's life. Eliminating suicide and homicide also increases life expectancy more than does eliminating such chronic diseases as atherosclerosis, the deaths from which are concentrated in the oldest years.[31]

Another way to examine these changes is to show the gain in life expectancy at birth for those who actually died from the particular cause. This produces very different life expectancies, shown in Table 6–9, for the same causes in Table 6–8 but with different rankings. Thus, the largest gains in life expectancy result if we eliminate certain causes originating in the perinatal period because saving infants allows them to live out the entire average life expectancy for the population as a whole. Much the same is true of congenital anomalies, but since they kill not just in early infancy but also childhood, the gains in life expectancy from eliminating them are less than for perinatal conditions. Similarly, large shares of deaths from homicide, suicide, and accidents occur among persons who would have lived an average of 30 or 40 years longer had they not been struck down, so eliminating those causes for persons who actually died from them also adds significantly to life expectancy. On the other hand, when we eliminate the causes concentrated in late adulthood, the gains are comparatively small. This applies to cancer, heart disease, diabetes, and stroke, for example, along with other degenerative diseases. The smallest gain for the population as a whole is from eliminating atherosclerosis, since it takes almost all its toll when persons are old enough to die soon from other causes.

Gains in eliminating certain causes vary widely by race and sex (see Table 6–8). Since black males have especially high death rates for so many causes, eliminating the most virulent ones yields more gain than for any other group. Those causes include cancer, especially of the respiratory and digestive systems; nonvehicular accidents; conditions originating in the perinatal period; pneumonia and influenza; homicide; and chronic liver disease and cirrhosis. But because death rates among black males are relatively high from many causes, eliminating ischemic heart disease creates a relatively modest gain in life expectancy. Blacks of both sexes experience larger gains than whites from eliminating hypertensive heart disease, cancer of the digestive and genital organs, nonvehicular accidents, stroke, conditions of the perinatal period, pneumonia and influenza, homicide, chronic liver disease and cirrhosis, diabetes, and infectious diseases, especially septicemia. These gains reflect earlier observations about causes that kill blacks and conditions that hinder disease prevention and treatment.

Differential gains in life expectancy by sex show that black females add more years from eliminating heart disease than do white females or males of either sex. Women also gain more than men from eliminating certain forms of cancer, especially breast cancer, and from eliminating stroke, diabetes, septicemia, and atherosclerosis. Furthermore, there are a few differences in this pattern according to race, but while the number of years gained may differ considerably, the sex

Table 6-9

Years Added to Life Expectancy by Eliminating Selected Causes of Death for Persons Who Would Have Died from Each Cause, by Race and Sex, 1979-81

Cause	All Races	White		Black	
		Male	Female	Male	Female
Certain conditions of perinatal period	74.28	71.18	78.76	64.99	73.72
Congenital anomalies	60.32	58.60	62.76	54.23	61.11
Homicide	39.17	35.90	39.97	32.81	39.07
Suicide	30.42	27.85	32.79	29.60	34.79
Accidents	30.26	31.49	25.19	27.77	25.73
Motor vehicle	38.31	37.09	38.88	30.26	36.30
Other	23.78	25.64	17.00	25.76	22.16
Chronic liver disease and cirrhosis	21.04	18.36	21.08	21.48	25.68
Infectious and parasitic diseases	15.46	14.27	14.23	16.44	26.74
Tuberculosis	14.69	11.27	12.33	15.78	17.07
Septicemia	12.01	11.38	11.04	13.36	13.26
Malignant neoplasms	15.36	13.67	16.27	14.99	17.30
Leukemia	16.03	14.88	16.37	16.56	18.34
Breast	16.02	8.92	17.65	10.07	18.44
Respiratory organs	15.49	13.35	17.23	14.62	17.89
Digestive organs	12.63	11.68	12.47	13.23	14.06
Genital organs	12.16	8.41	16.31	9.14	16.83
Urinary organs	11.97	10.74	11.78	12.19	13.42
Diseases of heart	14.06	14.16	12.43	15.90	15.80
Ischemic	12.44	12.99	10.74	13.54	12.91
Hypertensive	11.02	11.11	9.17	13.88	13.25
Dibetes mellitus	11.88	11.31	11.51	13.36	13.67
Chronic obstructive pulmonary diseases	11.69	9.88	12.90	12.29	16.82
Cerebrovascular diseases	9.23	8.74	8.74	11.87	11.91
Pneumonia and influenza	9.06	8.57	7.91	13.96	13.00
Atherosclerosis	6.09	6.12	5.71	7.60	7.52

Source: National Center for Health Statistics, Lester R. Curtin and Robert J. Armstrong, "United States Life Tables Eliminating Certain Causes of Death," U.S. Decennial Life Tables for 1979-81, Vol. I, No. 2 (Washington, DC: Government Printing Office, 1988), Table E.

variations between blacks and whites are small. Males generally gain more years of life expectancy from eliminating most causes; however, this is so because they have higher death rates than females for most of them.

Years of Potential Life Lost

These and other causes have not actually been eliminated, of course, so they continue to exact a measurable cost in potential years of life lost to each cause separately and to all collectively. This loss is *premature mortality* if it is applied to the population under 65; it can be expressed as either the annual total number of years lost or the number lost per 1,000 population. The latter measure is preferable in comparing various population segments because it controls for size, although it is worth noting that in 1985–87 all causes of death together cost an annual average of over 12 million years of life lost under age 65. Cancer alone was responsible for the loss of 1.8 million years, heart disease for another 1.6 million.[32] Thus, although hypothetically eliminating major causes of death may add a relatively small amount to life expectancy for the average person, any significant reduction in death rates for those causes results in cumulated annual savings of hundreds, thousands, or even millions of years of life. Ultimately, this remains a hypothetical situation, since most causes may never actually be eliminated, although the death toll for many can be reduced.

The average annual number of years of life lost per 1,000 population among black and white males and females in 1985–87 is shown in Table 6–10, which includes 11 major causes and several subcategories. The largest number of years lost to the population as a whole is from accidents, especially vehicular accidents, followed by cancer and heart disease. Suicide, homicide, and HIV infections follow in that order and reflect the amount of human resource that preventable causes cost the nation. In fact, if we account for deaths from heart disease and cancer attributable to cigarette smoking, then violence, smoking, and HIV infections cost about 25 years of life annually per 1,000 population under 65, or a total of over 5 million years.

Years of potential life lost per 1,000 blacks of both sexes are about twice the number lost per 1,000 whites.[33] The losses are greatest among black males for all causes in Table 6–10, except motor vehicle accidents and suicide; losses among white males are greater for them. Furthermore, while accidents cause the largest loss of potential life among men of both races, the greatest loss among women of both races is to malignant neoplasms, especially breast cancer. Heart disease is second among white males, but homicide has that distinction among blacks. In fact, the racial difference for homicide is larger than for any other cause. Accidents are in second place among white women, while heart disease is second for black women. In sex comparisons, white women have smaller losses of potential life than white men for all causes except diabetes and, of course, breast cancer. Black women have smaller losses than black men for all causes except breast cancer. These patterns reflect differential death rates by race and sex and show that the nation is losing a disproportionate number of black males to premature mortality, although the loss of black females is also far above the national average and somewhat higher among white males. The reason is the low death rate among white women, whose favorable mortality situation heavily influences the national statistical norm against which other groups compare either relatively unfavorably or very poorly.

ACTIVE LIFE EXPECTANCY

When causes of death are eliminated or, more realistically, when death rates from premature mortality are reduced, those reductions may or may not add to

Table 6–10
Years of Potential Life Lost Annually Per 1,000 Population Under 65 Years of
Age for Selected Causes of Death, by Race and Sex, 1985–87

Cause	All Races	White		Black	
		Male	Female	Male	Female
All causes	56.5	66.8	35.0	130.1	70.2
Accidents	10.9	16.0	5.4	19.3	6.3
Motor vehicle	6.8	10.0	3.7	9.2	2.9
Other	4.1	6.0	1.7	10.1	3.4
Malignant neoplasms	8.6	8.7	8.3	11.1	9.5
Respiratory organs	2.0	2.6	1.4	3.7	1.4
Breast (female)	1.1	...	2.1	...	2.5
Colorectal	0.6	0.7	0.6	0.8	0.7
Prostate	0.1	0.2	...	0.3	...
Diseases of heart	7.3	10.0	3.6	15.2	8.4
Ischemic	4.0	5.2	1.8	6.4	3.1
Suicide	3.1	5.4	1.4	3.4	0.7
Homicide	3.1	2.7	1.0	18.3	4.3
HIV infections[a]	1.7	2.5	0.2	7.1	1.7
Cerebrovascular diseases	1.2	1.0	0.9	2.9	2.4
Chronic liver disease and cirrhosis	1.1	1.4	0.6	2.9	1.4
Pneumonia and influenza	0.8	0.8	0.5	2.5	1.4
Chronic obstructive pulmonary diseases	0.6	0.6	0.5	1.2	0.7
Diabetes mellitus	0.6	0.6	0.4	1.1	1.0

Source: National Center for Health Statistics, Health, United States, 1989
(Washington, DC: Government Printing Office, 1990), Table 25.

[a]For 1987 only.

the survivors' quality of life. Consequently, there is a substantial qualitative difference between life expectancy and *active life expectancy*. The latter concept is also closely related to *years of functional health*, especially among the elderly.[34] Part of the distinction between life expectancy and active life expectancy relates to when the extra years of life are saved. Between ages 15 and 65 persons have already used significant resources and generally are or soon will be making their largest economic and other contributions.[35] Therefore, losses in the ages 15–65 are more strategic to a society than are losses among those elderly who are no longer productive or even losses among very young children in whom the society has not yet invested heavily. Those children have productive potential that the elderly cannot have, however, which is why the preceding section focused on potential years of life lost among persons under 65.

The age cutoff 65 and over is also the main reason why cardiovascular diseases have a comparatively small role in causing lost years because the effects of those causes are heavily concentrated in the oldest ages. Therefore, while heart disease and stroke are major killers, they are not responsible for nearly as large a loss in productive capacity as one might assume from their ranking as causes of death. In turn, while efforts to prevent and treat heart disease are laudable and defensible, society might make a more rational long-range investment in the productive capacity of its people by devoting more resources to reducing death from violence, cancer, and especially causes among children.[36] The same applies increasingly to HIV/AIDS, as the death rate for those causes continues to soar. In fact, the HIV/AIDS epidemic is getting so large that it severely strains the nation's health-care system, not only because of financial costs per patient, but by diverting human and other resources and in its psychological impact on medical personnel, families, and persons with AIDS themselves.[37] In this sense, efforts to reduce violence, to discover and reduce various preconditions of cancer, to lower infant mortality, and to prevent and cure AIDS represent an investment not just in longer life expectancy, but in longer active life expectancy. It is also important to remember, however, that the payoffs to society from making these several kinds of investments depend on their likelihood of success relative to their cost.

Societal interests are closely related to personal and family interests in this context. Increases in life expectancy, especially if they increase the percentage of people in the oldest ages, also may raise the incidence of dependency, illness, disability, and despondency, which can become severe burdens for elderly persons themselves and for their caregivers. This should not be applied too broadly because most elderly Americans are not seriously disabled, though additional efforts to raise life expectancy should not increase the share in poor health, but should aim for longer periods of good health and vigor.[38] Recent trends show that life expectancy has increased in the ages when chronic disabling illness is most likely, so longer life expectancy often does mean more years with health problems. For example, Sidney Katz and others found that because of such disabilities the age group 65–74 had an average of ten years of active life expectancy, or functional well-being, while the age group 85 and over had less than three years.[39] At the same time, there has been only a slight increase in years of life expectancy with a disabling illness severe enough to confine a person to bed. Therefore, despite the increase in years with chronic disabling illness, it does not appear that "advances in medical science are simply enabling us to spend increasing proportions of our lives as bed-ridden dependents."[40]

The need to focus on active life expectancy involves not just medical technology and health care in general, but their availability to various groups. For example, even though the death rate for coronary heart disease has fallen substantially, it still causes relatively high rates of morbidity, disability, and premature death among persons who rank low on socioeconomic measures, especially education. They are least likely to change diet, smoking, exercise, and other health habits associated with death rates for this cause and are, therefore, most likely to be its victims.[41] The disadvantaged are also likely to have shorter active life expectancy than more affluent groups. Consequently, efforts to prolong life expectancy should account for factors that determine whether those added years of life are actually worth living. Some of those factors are genetic, and while persons may not be strongly predisposed to inherit their parents' longevity, there does seem to be a strong correlation between the frailty of parents and children.[42] Even strategic genetic predispositions are not invariable constraints, however, and can be manipulated by environmental factors that affect actual life expectancy. The influence of socioeconomic conditions on life expectancy reflects many such factors, and unless the health-care system can manipulate them more equitably and improve the prospects of active longevity regardless of socioeconomic status but still contain costs, the system is likely to come under more bureaucratic regulation. That change may or may not contribute to universal well-being and longer periods of functional life that the health-care system presumably seeks.[43]

Many lifesaving technologies can sustain or even revive life, and to the extent they actually do so, the technologies increase life expectancy and change life table statistics, especially for the elderly.[44] We focus here on the elderly because, compared with younger people, they are at greater risk of life-threatening illnesses, are more prone to chronic conditions and disabilities, have a higher probability of mental impairments that keep them from making decisions about their own health care, and are more likely to suffer comorbidity, or several diseases simultaneously. Therefore, lifesaving technologies are most likely to be used to extend life expectancy of the elderly, thus consuming a large share of public health-care finances and re-emphasizing questions about active life expectancy.[45]

"Life-sustaining technologies are drugs, medical devices, or procedures that can keep individuals alive who would otherwise die within a foreseeable, but usually uncertain, time period."[46] Among others, they include the following: (1) *Cardiopulmonary resuscitation* (CPR) restores heartbeat and maintains breathing and blood flow after cardiac or respiratory arrest. It can be applied to virtually anyone whose heart stops beating, but the elderly are the most likely beneficiaries. (2) *Mechanical ventilation* inflates and deflates the lungs and regulates the exchange of blood gases, and while these technologies can be used to save lives at any age, the procedures among the elderly are especially likely to be prolonged or chronic, often preserving biological life after its other attributes are gone. (3) *Renal dialysis* maintains the chemical balance of the blood when the kidneys have failed for persons with either acute renal failure or chronic "end-stage renal disease." It is an expensive procedure, largely publicly funded. (4) *Nutritional support and hydration* provides nourishment and fluids artificially (e.g., tube feeding). These procedures are used with persons who cannot take food and fluids by mouth or digest and absorb them, and frequently they are employed with the elderly who are terminally ill. (5) *Antibiotics* are used for any life-threatening infection caused by bacteria or other microorganisms. They too save lives among persons of any age but are particularly crucial among those elderly whose immune systems are impaired because of illness or the aging process itself. Antibiotics can cure an underlying life-threatening condition, whereas the other technologies can only maintain life while a condition persists.[47]

These technologies have variable lifesaving potential, depending on whether a person is in a life-threatening crisis, is chronically ill with one or more conditions, has impaired functional capacity, or is terminally ill and no known treatment can stop the progression of a fatal disease.[48] While a small share of persons 65 and over is in these situations at any given time, a very large share of persons in them is 65 and over. The technologies can alter significantly the timing and circumstances of death, and most questions that surround their use involve the quality of life that is preserved and, by implication, the quality of death. Do the technologies prolong life or prolong dying? Is the result worth the cost? The questions are subjective and individualized and the answers lie with religion, philosophy, ethics, morals, personal conviction, and other areas outside the scope of social science and, therefore, outside the scope of this book. It is important to understand, however, that efforts to extend life expectancy, especially for the elderly, involve fundamental concepts such as active life expectancy and an acceptable quality of life. Therefore, societies with sophisticated lifesaving technologies have to decide what those concepts mean in particular sociocultural settings, how the meanings should affect criteria for rationing the technologies, and how and for whom life expectancy will or will not be extended.

HEALTH GOALS AND LIFE EXPECTANCY

If active life expectancy is to mean more than keeping a larger number of persons alive longer, the nation's health goals should address not only mortality, but overall physical well-being. That was the intent in 1979 when the Surgeon General articulated a range of goals to do more than just reduce mortality.[49] The principal goals were to improve the mortality and health levels of particular age groups and to extend life expectancy, and they were to be achieved by 1990. Some were mentioned previously, but summarized briefly, the goals were as follows.

Among those aged 0–1 health improvements were to lower infant mortality 35 percent to fewer than 9 deaths of infants per 1,000 live births. In 1990 the rate among whites was well below that level, but among blacks it was much higher.

Among children aged 1–14 health and childhood development were to lower the death rate at least 20 percent to fewer than 34 deaths per 100,000. In 1990 the death rate among white children was substantially below that level, the one among black children significantly above it.

Among adolescents and young adults aged 15–24 better health and health habits were to bring the death rate down at least 20 percent to fewer than 93.0 deaths per 100,000. Once again, in 1990 whites were below that level, though not by much, while blacks remained significantly above it.

Among adults aged 25–64 health practices and levels were to reduce the death rate at least 25 percent to fewer than 400 deaths per 100,000. In 1990 whites were below that level, especially in the ages 25–49, but the goal was achieved for blacks only in the ages 25–34. In the ages 25–64 collectively the goal was achieved for whites but not for blacks.

Among adults 65 and over improvements in health and quality of life were to reduce death rates and to reduce by 20 percent the number of days of restricted activity due to acute and chronic conditions, to fewer than 30 days per year. In addition, the number of bed-disability days due to various conditions would fall 20 percent to fewer than 12 days per year. These actually are indicators of active life expectancy. In 1990 the white population had become sufficiently healthy to meet the goals but the black population had not.[50]

In general, the goals were realistic for white Americans, given where they already stood when the goals were articulated and the average levels of health care and other lifesaving facilities and knowledge available to them, although many individuals still fell far below the minimal expectations for 1990. The goals proved largely unrealistic for black Americans as a whole because many lost ground socioeconomically in the 1980s; their quality and quantity of health care and other conditions that contribute to longevity either did not improve or deteriorated. This was more true for males than females, and for black males as a whole most of the 1990 goals were far above actual achievements.

In 1990 Louis Sullivan, Secretary of Health and Human Services, stated the next decade's health goals in *Healthy People 2000: National Health Promotion and Disease Prevention Objectives.*[51] He charted the progress made by 1990 and identified areas where the earlier goals were not met and, therefore, where additional efforts were needed. The new goals emphasize care for persons who are ill or disabled and a wide range of preventive measures to minimize the number who become ill or disabled. Based on this rationale, *Healthy People 2000* states three major objectives:

1. To increase not just life expectancy, but the years of healthy life expectancy, largely by helping individuals control ten factors that contribute to death and disability. The most crucial are "poor diet, lack of prenatal care, infrequent exercise, the use of tobacco, alcohol and drug abuse, and the failure to use seat belts." Better control of these factors alone "could prevent between 40 and 70 percent of all premature deaths, a third of all cases of acute disability, and two thirds of all cases of chronic disability."[52]

2. To reduce differences in health among various groups, especially by providing better health care to persons most at risk of illness and injury. This goal means more attention to the poor and minority groups so as to reduce present gaps in well-being and active life expectancy.[53] It also implies the need for a national health-care system.

3. To focus more on disease prevention so as to reduce the percentage of all expenditures devoted to treating avoidable illnesses and disabilities. Part of the goal is to improve the health of Americans, but part is to contain costs. According to Sullivan the 1988 health-care costs of $530 billion will increase to $1.5 trillion

by 2000, unless much more effective preventive measures are implemented. He cites decreased death rates for heart disease and stroke as results of success in reducing high blood pressure, cigarette smoking, and high blood serum choles-terol and dietary fat. Sullivan also says the reduced death rate for motor vehicle accidents reflects less driving while impaired by alcohol and drugs and more use of seat belts.[54]   *from Sec of Health/Human Service*

In sum, the health goals for 2000, intended to increase active life expectancy at all ages, focus more on prevention than treatment, though better treatment also is necessary, especially for groups that receive too little now. In many cases we know what preventive measures are needed, so the task is to inform people and provide better access to preventive mechanisms and facilities. That will cost more money, but much less than treating diseases and injuries that could have been avoided. In other cases, however, prevention requires comprehensive so-cioeconomic changes to reduce deaths from causes such as homicide, suicide, and drug and alcohol abuse; those efforts will be very expensive. Furthermore, significant changes in risky habits among a large share of the population may require far-reaching social restructuring, and that usually has high costs, whether financial, psychosocial, or political.

TRENDS IN LIFE EXPECTANCY

Several methodological difficulties, discussed in Chapter 1, affect efforts to trace trends in life expectancy. Until 1933, for instance, the registration data on which life expectancy is based were reported for registration-area states rather than the entire nation. In addition, early data, first gathered systematically for 1900, contain serious reporting errors, though even now data on life expectancy are most accurate in a census year (e.g., 1980, 1990), whereas accuracy dimin-ishes as population estimates (e.g., 1988, 1989) grow more remote from the enumeration on which they are based. Racial categories also were vaguely defined until 1970, when life expectancy was first reported specifically for blacks, al-though at times data were reported for ''Negroes.'' The ''all other'' or ''non-white'' categories did consist overwhelmingly of black persons for several decades: in 1900, for example, 96 percent of ''nonwhite'' races was black, so data for nonwhites are largely for blacks. By 1970, however, other racial groups had increased as a proportion of the total and blacks were down to 89 percent. Therefore, during part of the period prior to 1970 racial classifications became increasingly problematic in tracing life expectancy trends. In 1962 and 1963 the figures by race also excluded the data for New Jersey, and in 1972 the national figures were based on a 50 percent sample of deaths. The trends also are subject to age misreporting, especially in the base population data and age at death among the very old.

Despite the problems, trends in life expectancy are important enough for a look at the early patterns, followed by a more thorough examination of recent trends. Early curves, however, are approximations of what occurred, not highly accurate detailed portrayals, and even later ones are more accurate approximations rather than perfect reflections of reality.

Long-Term Trends

Annual changes in life expectancy at birth for the total population appear in Figure 6–1. The curve shows large year-to-year variations until about 1945 and much less spectacular annual changes thereafter. It also shows that increases in life expectancy were small and easily reversed in the first decade of the century and that they became fairly substantial, though sporadic, until the influenza epidemic caused a drastic decrease in 1918. The next few years also saw spectacular increases in life expectancy because apparently the epidemic "eliminated a large number of weak persons leaving an unusually hardy remainder," so the increases "in the 1910–20 decade were larger than would normally have been expected, and those for the 1920–30 decade were smaller."[55] Life expectancy then rose substantially, though still sporadically, until about 1949–50; it increased slowly until 1954, remained almost unchanged until the late 1960s, and then began a steady increase until about 1980, after which progress was slow. These changes correspond closely to variations in the age-adjusted death rate, discussed in Chapter 2.

The major factor in raising life expectancy from 47.3 years in 1900 to 68.2 years in 1950 was the impressive control gained over many infectious and parasitic diseases by vaccines and antibiotics. Widespread use of these death controls also reduced the wild year-to-year swings in death rates and life expectancy. Increases in life expectancy in the 1970s, however, following a virtual hiatus in the 1960s, resulted largely from reductions in general mortality, but especially deaths from heart disease, stroke, and motor vehicle accidents.[56] Therefore, reasons for rising life expectancy early in the century represent a different stage in the epidemiologic transition than reasons late in the century. Whatever the cause, the rise in life expectancy from 47.3 years in 1900 to 74.9 years in 1988 represents an increase of 27.6 years, or 58 percent.

Table 6–11 shows variations in life expectancy by sex and race, but in order to minimize annual fluctuations and data inaccuracies, especially for blacks, it is based on life expectancy at birth for 1900–02, several three-year decennial periods, and a final three-year period in the late 1980s. Life expectancy improved significantly for blacks and whites, males and females, during the entire period, but blacks had the largest gains, partly because their life expectancy was so low at the turn of the century. Thus, between 1900–02 and 1986–88 various groups gained as follows: black females, 38.5 years (110 percent); black males, 32.6 years (100 percent); white females, 27.8 years (54 percent); and white males, 23.9 years (50 percent). Furthermore, most groups followed the national pattern in which life expectancy rose sharply from the turn of the century until 1955 or so, leveled out until the late 1960s, rose again in the 1970s, and nearly leveled out in the 1980s. There were some variations, however, and enough females

Figure 6–1
Years of Life Expectancy at Birth, 1900–88

LIFE EXPECTANCY

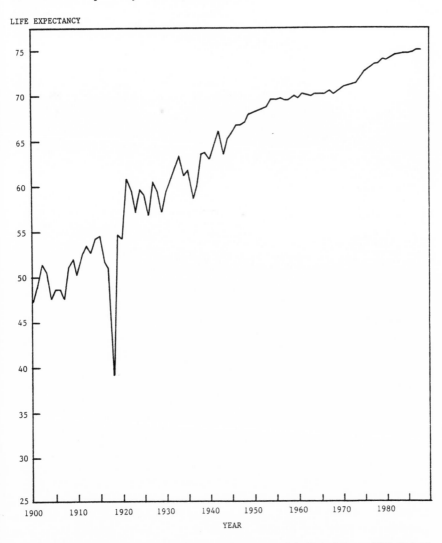

YEAR

Source: National Center for Health Statistics, *Vital Statistics of the United States, 1988*, Vol. II, *Mortality*, Part A (Washington, DC: Government Printing Office, 1991), Sec. 6, Table 5.

Table 6–11
Years of Life Expectancy at Birth for Selected Periods, by Race and Sex, 1900–02 to 1986–88

Time Period	White		Black	
	Male	Female	Male	Female
1900–02	48.23	51.08	32.54	35.04
1909–11	50.23	53.62	34.05	37.67
1919–21	56.34	58.53	47.14	46.92
1929–31	59.12	62.67	47.55	49.51
1939–41	62.81	67.29	52.26	55.56
1949–51	66.31	72.03	58.91[a]	62.70[a]
1959–61	67.55	74.19	61.48[a]	66.47[a]
1969–71	67.94	75.49	60.00	68.32
1979–81	70.82	78.22	64.10	72.88
1986–88	72.17	78.87	65.10	73.50

Source: National Center for Health Statistics, Robert J. Armstrong, "Some Trends and Comparisons of United States Life Table Data: 1900–1981," U.S. Decennial Life Tables for 1979–81, Vol. I, No. 4 (Washington, DC: Government Printing Office, 1987), Table A; NCHS, Vital Statistics of the United States, 1988, Vol. II, Mortality, Part A (Washington, DC: Government Printing Office, 1991), Sec. 6, Table 5.

[a]All races other than white.

died in the influenza epidemic of 1917–18 so that in 1919–21 life expectancy among black males was actually higher than among black females. Even though life expectancy of white females exceeded that of white males in every decade, the gap was smallest in 1919–21.[57] In most other decades, life expectancy of females of each race increased faster than that of males, although the situation reversed for both races in 1979–81 and 1986–88.

These patterns appear in the number of years by which life expectancy of females exceeds that of males, separately by race, and the number of years by which life expectancy of whites exceeds that of blacks, separately by sex (see Table 6–12). Long-term trends show a narrowing racial gap, though that progress also has slowed recently. Thus, in 1900–02 at birth white males had 15.7 more years of life expectancy than black males, but by 1986–87 the gap had narrowed to 7.1 years, though it was even smaller in 1979–81. Among females the racial gap narrowed from 16.0 years in 1900–02 to 5.4 years in 1986–88, though for females too the gap was slightly smaller in 1979–81. Thus, while life expectancy of blacks improved faster than that of whites over the long run, in the 1980s that was not true. The trends also show a substantial increase in the sex gap in life expectancy, apart from the sizable convergence in 1919–21, although the increase was slower after 1969–71 than it had been earlier. Among whites only

Table 6–12
Excess Years of Female Over Male Life Expectancy at Birth, by Race, and Excess of White Over Black, by Sex, 1900–02 to 1986–88

Time Period	Excess of Female Over Male		Excess of White Over Black	
	White	Black	Male	Female
1900–02	2.85	2.50	15.69	16.04
1909–11	3.39	3.62	16.18	15.95
1919–21	2.19	-0.22	9.20	11.61
1929–31	3.55	1.96	11.57	13.16
1939–41	4.48	3.30	10.55	11.73
1949–51	5.72	3.79[a]	7.40[a]	9.33[a]
1959–61	6.64	4.99[a]	6.07[a]	7.72[a]
1969–71	7.55	8.32	7.94	7.17
1979–81	7.40	8.78	6.72	5.34
1986–88	7.70	8.40	7.07	5.37

Source: National Center for Health Statistics, Robert J. Armstrong, "Some Trends and Comparisons of United States Life Table Data: 1900–1981," U.S. Decennial Life Tables for 1979–81, Vol. I, No. 4 (Washington, DC: Government Printing Office, 1987), Table B; NCHS, Vital Statistics of the United States, 1988, Vol. II, Mortality, Part A (Washington, DC: Government Printing Office, 1991), Sec. 6, Table 5.

[a]All races other than white.

2.8 years separated the sexes in 1900–02, but by 1986–88 that distance had grown to 7.7 years. Among blacks the sex gap was 2.5 years in 1900–02, 8.4 years in 1986–88, though it was even larger in 1979–81. For whites the sex gap widened gradually over most of the century, while for blacks the principal widening took place largely after 1960, as infectious and parasitic diseases and other sex-neutral causes of death diminished in importance and heart disease and other degenerative diseases, which take a much higher toll of males than females, became proportionately more important.

Life Expectancy at Various Ages. The major long-term improvements in years of life expectancy have come at birth because much larger percentages of newborns remain alive now than in the past, although significant improvements occurred at every other age as well because lifesaving technologies were applied across the age spectrum. In one sense those improvements grow smaller with advancing age, and while in 1985–87 a newborn white male could expect to live 23.8 years longer than in 1900–02, the life expectancy of a white man aged 65 increased only 3.3 years during that period (see Table 6–13). The percentage improvements, however, do not diminish consistently with age, and while life expectancy at birth for white males rose 49 percent between 1900–02 and 1985–87, at no age did the proportional improvement fall below 25 percent. At age 65 the percentage increase rose, and even though a white man aged 80 could

3

~ in Life Expectancy, by Age, Race, and Sex, 1900–02 to 1985–87

Age	Increase in Years				Percent Increase			
	White		Black		White		Black	
	Male	Female	Male	Female	Male	Female	Male	Female
0-1	23.8	27.7	32.7	38.5	49.4	54.2	100.6	110.0
1-5	17.1	22.0	22.7	30.3	31.3	39.0	53.4	69.7
5-10	13.5	18.6	16.7	24.0	24.8	33.2	37.0	52.2
10-15	12.4	17.4	15.0	22.1	24.5	33.3	35.8	51.4
15-20	11.8	16.9	13.7	20.4	25.5	35.4	35.8	51.3
20-25	11.2	16.1	12.2	15.4	26.5	36.8	34.8	41.7
25-30	10.3	14.9	7.8	16.6	26.8	37.2	24.2	49.0
30-35	9.3	13.7	9.2	15.5	26.6	37.6	31.4	49.2
35-40	8.2	12.5	8.1	13.6	26.1	38.1	30.9	49.5
40-45	7.2	11.3	7.2	12.2	26.0	38.7	31.2	50.0
45-50	6.3	10.3	6.3	10.8	26.0	40.4	31.3	50.5
50-55	5.2	9.3	5.4	9.3	25.0	42.5	31.2	49.7
55-60	4.5	8.4	4.6	8.1	25.9	45.7	31.3	50.9
60-65	3.8	7.4	3.5	6.7	26.4	48.7	27.8	49.3
65-70	3.3	6.5	3.0	5.6	28.7	53.3	28.8	49.1
70-75	2.7	5.5	2.5	4.3	30.0	57.3	30.1	44.8
75-80	2.3	4.5	2.1	3.2	33.8	61.6	31.8	40.5
80-85	1.8	3.3	1.7	2.1	35.3	60.0	33.3	32.3

Source: National Center for Health Statistics, Vital Statistics of the United States, 1985, 1986, 1987, Vol. II, Mortality, Part A (Washington, DC: Government Printing Office, 1988–90), various tables.

expect to live an average of only 1.8 years longer in 1985–87 than in 1900–02, that was a 35 percent increase.

Much the same pattern pertains for white women, except that at every age the number of years added to life expectancy between 1900–02 and 1985–87 was considerably larger for females than males. The same was true of percentage increases, which were so large among elderly white women that in all ages 70 and over life expectancy actually grew by a greater percentage than it did among infants. This reflects the unusually long life expectancy of white females regardless of age, contrasted with white males and blacks of both sexes.

Black males had larger increases in years of life expectancy than white males up to age 25, when the increase for blacks dropped sharply because of homicide rates. In all subsequent ages, however, the increase in years of longevity was about the same for black and white males. At most ages, except 25–30 and 65 and over, the increase in years among black males represents a larger percentage growth in longevity than the increase in years among white males, so there was some convergence during the century. In fact, life expectancy among white males at birth improved nearly 50 percent, but that among black males at birth rose 100 percent. While black males are still the least advantaged group as measured by life expectancy at various ages, the magnitude of the disadvantage relative to white males has diminished, especially for infants and children.

Black females aged 0–20 and 25–50 experienced larger increases than any other group in years of life expectancy. After age 50 the number of years added was somewhat smaller for black than white females, but it was larger than the increases among males of both races. In addition, life expectancy among black females increased by larger percentages than for any other race or sex group in all ages 0–65. Only at age 65 and over were the percentage increases larger for white women, though by a considerable margin. These data show the impressive extent to which the mortality and longevity situation has improved in the black female population since 1900–02, although given their death rates for several causes there is room for much more improvement.

Short-Term Trends

Inasmuch as data on life expectancy among blacks were first reported separately from other races in 1970, we can use annual figures to trace trends by race and sex since that year. Figure 6–2 shows the following trends in life expectancy at birth.

1. White females enjoyed far higher life expectancy than all other groups throughout the period, although there was a slight convergence with white males, especially after 1982.

2. Life expectancy of black females, which was below that of white males prior to 1955, was somewhat above by 1970 and the two continued to diverge until 1982. Then life expectancy among white males continued to increase slowly while that among black females remained static or even dropped slightly.

3. Life expectancy among black males rose fairly rapidly between 1970 and 1985, though with two brief reversals because of influenza epidemics. Then, however, their life expectancy actually declined from 65.6 years in 1985 to 64.9 years in 1988. Black females also experienced a decrease during that period, but it was smaller than for black males.

4. In the 1980s white males had the greatest improvement, but while the first half of the 1980s saw relatively rapid improvement among black males, it was followed by the reversals mentioned above. The 1980s were not especially kind to any group, however, for females of both races experienced no significant improvements during that time, and even the increases in life expectancy for males were comparatively small or compromised by reversals.

Slow progress in the 1980s suggests that the nature of health care, poverty, specific causes of death, lifestyles, and other realities in American society may have brought us close to the limit on life expectancy unless we are willing to implement the improved lifesaving efforts recommended in *Healthy People 2000.* Relatively static life expectancy curves reflect relatively little progress, or even regress, in trying to solve many problems that influence mortality rates and

Figure 6–2
Years of Life Expectancy, by Race and Sex, 1970–88

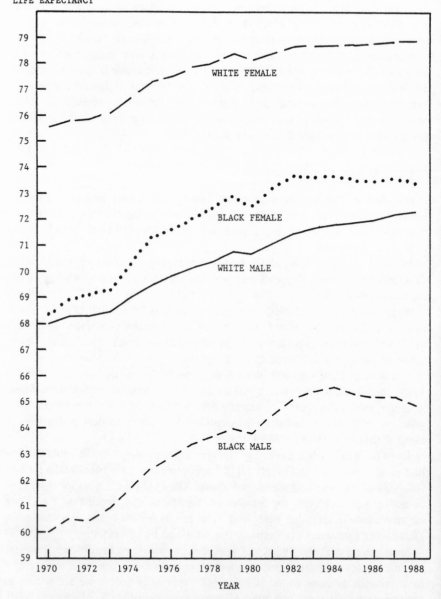

LIFE EXPECTANCY

WHITE FEMALE

BLACK FEMALE

WHITE MALE

BLACK MALE

YEAR

Source: National Center for Health Statistics, *Vital Statistics of the United States, 1988*, Vol. II, *Mortality*, Part A (Washington, DC: Government Printing Office, 1991), Sec. 6, Table 5.

longevity, either directly or indirectly. This is not to denigrate spectacular new medical technologies, steady reduction in the death rate for heart disease, research on ways to help people with AIDS and other diseases, or various preventive measures that reduce the virulence of many causes of death. Relatively static life expectancy does show, however, that most death controls are applied very unevenly across the population, and if over 30 million Americans are going to continue without health insurance and others are going to be without adequate medical care for various reasons, we must expect life expectancy to increase slowly for some groups and not at all for others, while some lose ground. We especially need to decrease infant mortality rates, at least to the average for developed societies, because conditions among infants are more important in determining whether persons will live long healthy lives than are any efforts to reduce death rates among the elderly. In fact, many recent spectacular medical advances (e.g., organ transplants, angioplasty, coronary-artery surgery) are directed largely to older persons, whose life expectancy is brief, while resources are scarce for many of the youngest vulnerable persons, who potentially have most of their lives before them. In a nation as wealthy, creative, and presumably humane as the United States, neglecting a sizable share of infants, children, adolescents, and young adults, with adverse consequences for life expectancy, is not a necessity but the result of choosing certain priorities over others. The relative lack of progress recently in raising life expectancy originates partly in the inequitable distribution of health care, both preventive and curative. That maldistribution, in turn, "is rooted in the incapacity of [our] political institutions to ensure that the burdens and benefits of [economic] adjustment [are] allocated fairly."[58] In practice, poor progress means that more of the responsibility for health care has shifted from federal funding to state and local areas that cannot ·or will not carry the burden.

In assessing this apparent lack of progress, we should recognize that the 1980s represent a short period in the history of mortality change and that there is room for optimism about future change. Indeed, the pace of improvement has been quite uneven, and Table 6–11 shows that while there was almost no progress in life expectancy for whites of either sex or for black males between 1950 and 1970, from 1970 to 1987 there were substantial gains: four years for white males, three years for white females, and over five years for blacks of both sexes. Therefore, with a longer perspective than just the 1980s, we might well expect the pace of improvement in life expectancy to accelerate once again as we actually do move to implement the goals stated in *Healthy People 2000*.

The narrowing sex differential in life expectancy in the late 1980s relates to changes in the cause of death, illustrated by those between 1987 and 1988. While age-adjusted death rates among males continue to be higher than those among females for each of 15 leading causes of death, between 1987 and 1988 the age-adjusted rate among females for all causes combined rose slightly, while that among males fell slightly. The two major causes of death that contributed to the narrowing sex mortality differential are cancer and heart disease: the age-adjusted death rate for cancer fell slightly among males and rose among females, while the rate for heart disease fell faster among males than females. Race is involved here too, and the small sex convergence in longevity during the late 1980s came from unchanged life expectancy among white females, combined with a drop of 0.2 year among black females, an increase of 0.1 year among white males, and a drop of 0.3 year among black males.[59]

Life Expectancy at Various Ages. Patterns of improvement in age-specific life expectancy between 1969–71 and 1985–87 are similar in some ways to those between 1900–02 and 1985–87, but different in others (see Table 6–14). In both periods and for blacks and whites of both sexes, the largest improvements in years of life expectancy took place among newborns and the number of years added to life decreased almost without exception in each succeeding age group. But the percentage changes in life expectancy show different patterns. Between 1969–71 and 1985–87 white males experienced a larger percentage increase in life expectancy at birth than in the ages 1–15. But then the percentage rose steadily with age until white men aged 65 had a proportional improvement in life expectancy more than twice that among infants, and while percentage improvements were more modest in some older ages, they were still well above the 6 percent for newborns. The pattern was similar for white females, although the percentage improvements in life expectancy remained below those of white males until age 70, after which they were greater.

The black population shows similar changes: years added to life expectancy were greater for newborns of both sexes than at any older age. In fact, among black men aged 75–85 and black women aged 80–85 the years of life expectancy actually decreased slightly between 1969–71 and 1985–87. The percentage changes, however, differ from those among whites. Among black males life expectancy at birth increased 9 percent; it fell for young children and then generally rose until age 45–50, only to fall again, most notably in the ages 75–85. The percentage improvement in life expectancy among black males was greater in the ages 30–60 than for newborns, but in other ages the youngest persons had the edge. Among black females the pattern was roughly similar, except that women aged 20–70 had higher percentage increases in life expectancy than did infants.

These patterns suggest that the impact of lifesaving conditions on life expectancy changed in several ways from the first half to the second half of the century. In the early years, when infectious and parasitic diseases still caused a large share of deaths, improvements in sanitation, nutrition, and medical care increased the longevity of infants more than other age groups. After most of those causes had declined in importance and as the population aged, attention directed to heart disease, cancer, stroke, and other degenerative causes in the older ages worked to the disproportionate advantage of middle-aged and elderly adults. Therefore, their life expectancy increased by larger percentages than that of infants and children and even young adults. In that sense the recent advances in lifesaving technology have contributed to the aging of the population. Typi-

Table 6-14
Increase in Life Expectancy, by Age, Race, and Sex, 1969-71 to 1985-87

Age	Increase in Years				Percent Increase			
	White		Black		White		Black	
	Male	Female	Male	Female	Male	Female	Male	Female
0-1	4.1	3.3	5.2	5.2	6.0	4.4	8.7	7.6
1-5	3.4	2.7	4.4	4.4	5.0	3.6	7.2	6.3
5-10	3.3	2.7	4.2	4.3	5.1	3.8	7.3	6.5
10-15	3.3	2.6	4.1	4.2	5.5	3.9	7.8	6.9
15-20	3.3	2.6	4.0	4.2	6.0	4.2	8.3	7.5
20-25	3.2	2.7	3.8	4.1	6.4	4.7	8.7	8.0
25-30	3.1	2.6	3.4	3.9	6.8	5.0	8.6	8.4
30-35	3.1	2.5	3.1	3.8	7.5	5.2	8.8	9.0
35-40	3.1	2.5	2.9	3.5	8.5	5.8	9.2	9.3
40-45	3.0	2.4	2.7	3.3	9.4	6.3	9.8	9.9
45-50	2.9	2.3	2.4	2.9	10.5	6.9	10.0	9.9
50-55	2.7	2.1	2.0	2.5	11.6	7.2	9.7	9.8
55-60	2.4	1.9	1.6	2.0	12.3	7.6	9.0	9.1
60-65	2.1	1.8	1.2	1.6	13.0	8.7	8.1	8.6
65-70	1.8	1.8	0.9	1.3	13.8	10.7	7.2	8.3
70-75	1.3	1.7	0.4	0.9	12.5	12.7	3.8	6.9
75-80	1.0	1.6	-0.1	0.2	12.3	15.7	-1.1	1.8
80-85	0.7	1.2	-0.6	-0.3	11.3	15.8	-8.1	-3.4

Source: National Center for Health Statistics, Vital Statistics of the
United States, 1985, 1986, 1987, Vol. II, Mortality, Part A
(Washington, DC: Government Printing Office, 1988-90), various tables.

cally, aging occurs because the birth rate drops, infants and children become a smaller percentage of the total population, and older persons born when birth rates were high automatically become a larger share. Their death rates fall, but not as much as those among infants, so while the drop adds to the number of elderly it may not increase their percentage, and aging of the population takes place in spite of, not because of, differential improvements in life expectancy.[60] Recently, however, the effects of mortality control have been felt more evenly across the age spectrum.[61] In fact, the greater percentage improvements in life expectancy among older persons than among infants, along with low birth rates, have helped age the population. This is more the case in the white population than the black, whose percentage increases in longevity at some older ages remain below those in infancy.

SOME PROJECTIONS

Efforts to project life expectancy are based on current mortality patterns by age, race, and sex, and on other assumptions about possible changes in those patterns, though we cannot foresee them all.[62] Therefore, the Census Bureau projects life expectancy based on three sets of assumptions that postulate (1) a significant drop in mortality ("low series"), (2) a significant rise in mortality ("high series"), and (3) the extrapolation of present mortality ("middle series"). Middle-series assumptions and projections that follow from them are used in this section, though only to suggest the likelihood of certain changes in life expectancy, not to imply certainty. The projections are for five-year periods from 1990 to 2030 and ten-year periods from 2030 to 2080, though the farther one moves into the future, the greater the probable error.

If middle-series assumptions about mortality prove accurate, life expectancy at birth among males will rise from 72.1 years in 1990 to 77.8 years in 2080, for an increase of 5.7 years, or 8 percent (see Figure 6–3). Life expectancy in the female population will rise from 79.0 years in 1990 to 84.7 years in 2080, also for an increase of 5.7 years, or 7 percent.[63] These are much smaller than the increases between 1900 and 1990, so the projection rationale assumes that spectacular savings in lives are largely over and that the future will see relatively modest improvements. Although that rationale could prove too conservative because of unanticipated developments in lifesaving technology and practices, especially for infants, it is fairly consistent with the mortality dynamics of a society well into the last stage of the epidemiologic transition.

Projected increases in life expectancy at age 65 suggest greater proportional improvement than at birth. Under middle-series assumptions life expectancy among males would rise from 15.0 years to 18.8 years, for an increase of only 3.8 years, but 25 percent. Longevity among females would increase from 19.5 years to 23.9 years, for an increase of 4.4 years, or 23 percent. These projected increases also are consistent with events late in the epidemiologic transition.

Current racial differences in longevity would be eliminated under middle-series assumptions. Thus, while 5.0 years of life expectancy at birth separated black and white males in 1990, by 2080 that difference is expected to disappear, giving both groups a life expectancy of 77.8 years. For females the difference of 4.6 years is also expected to disappear, leaving blacks and whites with a life expectancy of 84.7 years in 2080. This total convergence assumes that socioeconomic and other conditions that now preserve the racial difference in life expectancy also will disappear and that the nation will achieve egalitarian access to health care, disease and violence prevention, and other lifesaving measures. That optimism will prove unjustified if the socioeconomic separations of the 1980s and early 1990s persist, but possibly the necessary fundamental structural changes to eradicate them will be made. If so, the nation will have to undergo a major social revolution that has proven elusive so far.

The sex differential in life expectancy at birth, unlike the racial differential, shows no signs of disappearing under the middle-series assumptions for whites, and while 6.9 years of life expectancy at birth separated white males and females in 1990, by 2080 the sexes would still be 6.9 years apart. Given the tendency toward convergence between white males and females in the 1980s, however,

Figure 6–3
Projected Life Expectancy, by Race and Sex, 1990–2080

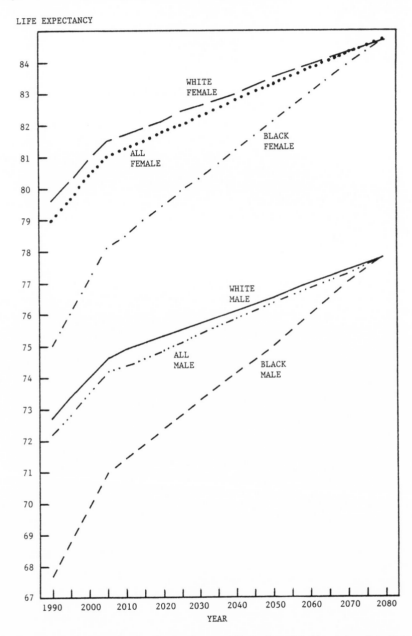

Source: U.S. Bureau of the Census, Gregory Spencer, "Projections of the Population of the United States, by Age, Sex, and Race: 1988–2080," *Current Population Reports*, Series P–25, No. 1018 (Washington, DC: Government Printing Office, January 1989), Table B–5.

the lack of convergence in the projections is questionable. The sex differential among blacks is projected to diminish, but by only 0.4 year between 1990 and 2080. A reduction is consistent with the small convergence that occurred in the 1980s, but the magnitude of the latter was larger than the one projected after 1990, and it too may prove unduly conservative.

Projections of life expectancy at age 65 also assume convergence by race but not sex, just as at birth. Thus, in 1990 elderly white men had 15.1 years of life expectancy remaining and black men had 13.9 years, but by 2080 both would have 18.8 years. In 1990 elderly white women had an average of 19.5 years left, black women 18.0 years, and both would have 23.9 years remaining in 2080. These changes would leave the life expectancy of white females 5.1 years higher than that of white males, which represents a divergence from 4.4 years in 1990. Black women aged 65 also would have 5.1 more years of life expectancy than black men, up from 4.1 years in 1990.

NOTES

1. John R. Weeks, *Population: An Introduction to Concepts and Issues,* 4th ed. (Belmont, CA: Wadsworth, 1989), p. 163. See also Henry S. Shryock and Jacob S. Siegel, *The Methods and Materials of Demography,* Vol. 2 (Washington, DC: Government Printing Office, 1973), Chap. 5; and A. H. Pollard, Farhat Yusuf, and G. N. Pollard, *Demographic Techniques,* 3d ed. (Sydney, Australia: Pergamon Press, 1990), Chap. 3.

2. National Center for Health Statistics, *Vital Statistics of the United States, 1987,* Vol. II, *Mortality,* Part B (Washington, DC: Government Printing Office, 1990), Sec. 6, p. 1. See also John A. Ross, "Life Tables," in *International Encyclopedia of Population,* Vol. 2, edited by John A. Ross (New York: Free Press, 1982), p. 421.

3. John Saunders, *Basic Demographic Measures* (Lanham, MD: University Press of America, 1988), p. 54.

4. Ansley J. Coale and Graziella Caselli, "Estimation of the Number of Persons at Advanced Ages from the Number of Deaths at Each Age in the Given Year and Adjacent Years," *Genus* 46:1 (January-June 1990), p. 1. See also Ansley J. Coale and Guang Guo, "Revised Regional Model Life Tables at Very Low Levels of Mortality," *Population Index* 55:4 (Winter 1989), pp. 613–643. Other methods for estimating life expectancy in the oldest ages are Shiro Horiuchi and Ansley J. Coale, "A Simple Equation for Estimating the Expectation of Life at Old Ages," *Population Studies* 36:2 (July 1982), pp. 317–326; and Samarendranath Mitra, "Estimating the Expectation of Life at Older Ages," *Population Studies* 38:2 (July 1984), pp. 313–319.

5. Robert D. Retherford, *The Changing Sex Differential in Mortality* (Westport, CT: Greenwood Press, 1975), p. 25.

6. Ansley J. Coale and Ellen Eliason Kisker, "Mortality Crossovers: Reality or Bad Data?" *Population Studies* 40:3 (November 1986), p. 401.

7. John H. Pollard, "On the Decomposition of Changes in Expectation of Life and Differentials in Life Expectancy," *Demography* 25:2 (May 1988), pp. 265, 274.

8. Eduardo E. Arriaga, "Measuring and Explaining the Change in Life Expectancies," *Demography* 21:1 (February 1984), p. 84.

9. Pollard et al., *Demographic Techniques,* p. 34.

10. A national agenda to narrow the racial mortality gap is U.S. Department of Health and Human Services, *Report of the Secretary's Task Force on Black and Minority Health* (Washington, DC: Government Printing Office, 1985).

11. NCHS, *Mortality*, Sec. 1, Table 9.

12. U.S. Bureau of the Census, Arlene F. Saluter, "Marital Status and Living Arrangements: March 1989," *Current Population Reports*, Series P–20, No. 445 (Washington, DC: Government Printing Office, 1990), Table 1. See also Paul E. Zopf, Jr., *America's Older Population* (Houston, TX: Cap and Gown Press, 1986), pp. 84–85.

13. U.S. Bureau of the Census, Paul C. Glick, "The Future of the American Family," *Current Population Reports*, Series P–23, No. 78 (Washington, DC: Government Printing Office, 1979), p. 1.

14. Ibid.

15. For some of these, see Helen Znaniecki Lopata, "The Widowed Family Member," in *Transitions of Aging*, edited by Nancy Datan and Nancy Lohman (New York: Academic Press, 1980), p. 107.

16. Ibid., p. 108.

17. Zopf, *America's Older Population*, p. 91.

18. For a method to overcome this problem, see David A. Swanson, "A State-Based Regression Model for Estimating Substate Life Expectancy," *Demography* 26:1 (February 1989), pp. 161–170.

19. National Center for Health Statistics, "State Life Tables," Florida, *U.S. Decennial Life Tables for 1979–81*, Vol. II, No. 10 (Washington, DC: Government Printing Office, 1985), Table 12; NCHS, Lester R. Curtin and Robert J. Armstrong, "United States Life Tables Eliminating Certain Causes of Death," *U.S. Decennial Life Tables for 1979–81*, Vol. I, No. 2 (Washington, DC: Government Printing Office, 1988), Table 3.

20. Verna M. Keith and David P. Smith, "The Current Differential in Black and White Life Expectancy," *Demography* 25:4 (November 1988), p. 625.

21. For the data, see Robert D. Grove and Alice M. Hetzel, *Vital Statistics Rates in the United States: 1940–1960* (Washington, DC: Government Printing Office, 1968), Table 51; and National Center for Health Statistics, *Vital Statistics of the United States, 1988*, Vol. II, *Mortality*, Part B (Washington, DC: Government Printing Office, 1991), Sec. 6, Table 1.

22. Keith and Smith, "Black and White Life Expectancy," p. 625.

23. Lloyd B. Potter and Omer R. Galle, "Residential and Racial Mortality Differentials in the South by Cause of Death," *Rural Sociology* 55:2 (Summer 1990), p. 234.

24. NCHS, *Mortality*, Part A, Sec. 6, Table 1.

25. A few such studies are Samuel H. Preston, *Mortality Patterns in National Populations with Special Reference to Recorded Causes of Death* (New York: Academic Press, 1976), Chap. 3; Nathan Keyfitz, "What Difference Would It Make If Cancer Were Eradicated? An Examination of the Taeuber Paradox," *Demography* 14:4 (November 1977), pp. 411–418; Kenneth G. Manton and Eric Stallard, *Recent Trends in Mortality Analysis* (New York: Academic Press, 1984), pp. 182–211; Josianne Duchene and Guillaume Wunsch, "From the Demographer's Cauldron: Single Decrement Life Tables and the Span of Life," *Genus* 44:3 (July-December 1988), pp. 1–17; and Keith and Smith, "Black and White Life Expectancy."

26. For the data, see NCHS, Curtin and Armstrong, "Eliminating Certain Causes of Death," Tables D, 22, 23, 24, 27, 28.

27. Preston, *Mortality Patterns*, p. 56.

28. Ibid.

29. NCHS, Curtin and Armstrong, "Eliminating Certain Causes of Death," p. 6.

30. Ibid., p. 7.

31. Ibid.

32. National Center for Health Statistics, *Health, United States, 1989* (Washington, DC: Government Printing Office, 1990), pp. 2, 125.

33. Ibid., p. 2.

34. For the concepts, see Sidney Katz, Laurence G. Branch, Michael H. Branson, Joseph H. Papsidero, John C. Beck, and David S. Greer, "Active Life Expectancy," *New England Journal of Medicine* 309:20 (November 17, 1983), pp. 1218–1224.

35. Preston, *Mortality Patterns*, p. 59.

36. Ibid., p. 60.

37. Sheldon H. Landesman, Harold M. Ginzberg, and Stanley H. Weiss, "The AIDS Epidemic," *New England Journal of Medicine* 312:8 (February 21, 1985), p. 523.

38. Edward L. Schneider and John D. Reed, Jr., "Life Extension," *New England Journal of Medicine* 312:8 (May 1, 1985), p. 1164.

39. Katz et al., "Active Life Expectancy," p. 1218.

40. Eileen M. Crimmins, Yasuhiko Saito, and Dominique Ingegneri, "Changes in Life Expectancy and Disability-Free Life Expectancy in the United States," *Population and Development Review* 15:2 (June 1989), p. 255.

41. Jeremiah Stamler, "Coronary Heart Disease: Doing the Right Things," *New England Journal of Medicine* 312:16 (April 19, 1985), p. 1054.

42. James W. Vaupel, "Inherited Frailty and Longevity," *Demography* 25:2 (May 1988), pp. 277, 281.

43. Arnold S. Relman, "Is Rationing Inevitable?" *New England Journal of Medicine* 322:25 (June 21, 1990), p. 1810.

44. James W. Vaupel and Anatoli I. Yashin, "Repeated Resuscitation: How Lifesaving Alters Life Tables," *Demography* 24:1 (February 1987), p. 123.

45. U.S. Congress, Office of Technology Assessment, *Life-Sustaining Technologies and the Elderly* (Washington, DC: Government Printing Office, 1987), pp. 5–6.

46. Ibid., p. 4.

47. The five technologies are from ibid.

48. Ibid., p. 6.

49. Assistant Secretary for Health and Surgeon General, *Healthy People: The Surgeon General's Report on Health Promotion and Disease Prevention* (Washington, DC: Government Printing Office, 1979).

50. For a summary of the goals, see NCHS, *Health, 1989*, p. 147. For the survey questions about health practices, see National Center for Health Statistics, Owen T. Thornberry, "Health Promotion Data for the 1990 Objectives," *Advance Data from Vital and Health Statistics* 126 (Washington, DC: Government Printing Office, September 1986).

51. Louis Sullivan, *Healthy People 2000: National Health Promotion and Disease Prevention Objectives* (Washington, DC: Government Printing Office, 1990).

52. Louis Sullivan, "Healthy People 2000," *New England Journal of Medicine* 323:15 (October 11, 1990), p. 1066.

53. Ibid.

54. Ibid., pp. 1066–1067.

55. National Center for Health Statistics, Robert J. Armstrong, "Some Trends and

Comparisons of United States Life Table Data: 1900–1981,'' *U.S. Decennial Life Tables for 1979–81*, Vol. I, No. 4 (Washington, DC: Government Printing Office, 1987), p. 2.

56. Ibid.

57. Ibid.

58. Robert B. Reich, ''What We Can Do,'' in *Crisis in American Institutions*, 7th ed., edited by Jerome H. Skolnick and Elliott Currie (Glenview, IL: Scott, Foresman, 1988), p. 594.

59. National Center for Health Statistics, ''Advance Report of Final Mortality Statistics, 1988,'' *Monthly Vital Statistics Report* 39:7 (Washington, DC: Government Printing Office, November 1990), pp. 4, 7–8.

60. U.S. Bureau of the Census, ''Demographic Aspects of Aging and the Older Population in the United States,'' *Current Population Reports*, Series P–23, No. 59 (Washington, DC: Government Printing Office, 1978), p. 4.

61. Pollard et al., *Demographic Techniques*, p. 126.

62. U.S. Bureau of the Census, Gregory Spencer, ''Projections of the Population of the United States, by Age, Sex, and Race: 1988 to 2080,'' *Current Population Reports*, Series P–25, No. 1018 (Washington, DC: Government Printing Office, 1989), p. 14.

63. For the data, see ibid., p. 153.

7 Conclusion

The principal focus of this book is twofold. First, it concerns present patterns of mortality, including general mortality and differentials among various groups, the mortality situation of the United States relative to other nations, infant mortality, and life expectancy. Second, it traces revolutionary mortality changes in the United States, including long-term reductions in general death rates and infant mortality rates, increases in life expectancy, changes in the relative importance of various causes of death, and fluctuations in mortality differentials among population components, especially males and females and blacks and whites. The book also accounts for certain social, economic, and other causes of mortality patterns and changes, and for some social, economic, and other results of those patterns and changes.

PERSPECTIVES ON MORTALITY

In the twentieth century the United States moved through the epidemiologic transition, from the stage when numerous communicable diseases were major causes of death, to the stage when degenerative illnesses claim a much larger majority of lives. That movement has greatly lengthened life expectancy, but not life span, which is supposedly an immutable species attribute but which may prove more flexible than is thought. Biological and environmental factors are involved in survival and longevity, and both are increasingly manipulable in ways that alter mortality patterns and life expectancy.

Measuring mortality encounters a variety of problems in data quality, which is affected by changes in the death-registration area, interpretations of cause of death, reporting and coding accuracy, and variable age profiles of different population segments. Some problems make for gaps in knowledge about mortality, but others can be solved or ameliorated by using numerous indexes ranging from simple to complex. Most indexes incorporate controls for age differences; they also are used to analyze population segments separately and thus to compare their respective mortality situations. The major indexes are the crude death rate, which is not controlled for age; the age-sex-specific death rate; the age-adjusted death rate, which is a standardized rate; the mortality ratio, which enables comparison between two populations; the infant mortality rate, components of which are the perinatal mortality ratio, the neonatal mortality rate, and the postneonatal mortality rate; the maternal mortality rate; the cause-specific death rate; life expectancy and the life table; and mortality differentials measured by most of the other indexes. Together, they constitute a sophisticated constellation of techniques to organize aggregate data derived from the individual death certificates though each index is only as good as those basic data.

GENERAL MORTALITY AND DIFFERENTIALS

The United States' mortality situation does not compare very favorably with those in most other developed nations; it is little better than those in some developing nations. This is largely because the United States is the only developed nation except South Africa that lacks a national health plan to provide universal health care, and because some groups of Americans have unusually high death rates. Black males fare especially badly as infants, boys, young men, and adults at nearly all other ages. In fact, one basic theme in the book is how poorly the health and mortality of black males compare with black females, whites of both sexes, Hispanics, and all other racial and ethnic groups.

Racial and ethnic mortality patterns vary widely by socioeconomic standing, and black males as a group have high death rates because as a group they are also inordinately affected by other problems. They endure relatively high stress levels on the job, have high rates of unemployment, and are much more likely than whites to be poor and unable to afford proper nutrition, preventive health care, and treatment, and to have risky lifestyles. Many black males and their families are caught between earning too much to qualify for Medicaid but too little to afford private health insurance. Even when they seek medical assistance it is mostly from white medical personnel, who may not understand the realities of black health and problems. Racial stereotypes also still influence parts of the health-care system, and blacks get poorer care and less attention than whites. Black females are subject to many of these same realities and the black female-headed family is especially vulnerable to conditions that produce illness, violence, and death of infants, children, and adults. Yet black females fare better than black males on virtually all measures of mortality, partly because of the biological protections of being female, but partly because of certain conditions in the sociocultural environment. Those conditions are so influential that when we hold them constant and compare blacks and whites of the same sex, most of the health and mortality differentials disappear. Therefore, they are not due in any significant degree to genetic or cultural differences, but largely to economic inequities and concomitant problems. For that reason, at least a third of black deaths are excessive—they could be prevented if blacks had health protection comparable to whites. The excess deaths include those from communicable diseases, degenerative illnesses, and violence, especially homicide.

In the ages 85 and over, a racial mortality crossover is reported in which the death rates of blacks are lower than those of whites. Age misreporting and other data inaccuracies play a part, but the crossover also appears at younger ages for certain causes of death, and at least part of it seems real. Thus, the oldest ages are populated by many frail whites who have been kept alive by sophisticated medical care, whereas many blacks in those ages are the survivors of poor medical and health conditions that eliminated their more vulnerable comrades at younger ages. Many elderly blacks have greater durability than many elderly whites, although various changes in the 1980s reduced the crossover effect.

Females of all ages generally have much lower mortality rates than males, and biological differences do play a significant role, though inevitably they interact with environmental forces. Variations in cigarette smoking, for example, explain part of the sex differential for several causes of death, though women have lost some of their health advantage as their smoking behavior and other circumstances have become more like those of men. Biological and environmental forces play different roles at different ages, and at some ages the mortality situations of the sexes are much more similar than at others; many variations diminish in old age, for example, though they rarely disappear. Women are more likely to seek care for an illness, and while they report more acute and chronic illnesses than men, women's maladies are generally less life threatening.

Other factors also correlate with significant mortality differentials. Married persons have lower mortality and morbidity rates than those who are single, widowed, or divorced, no matter what their age, race, or sex. Death rates are also lower at the higher levels of education, and as well-educated cohorts enter the ages 65 and over, death rates of the elderly should fall substantially, thus adding to their number and proportion and contributing to the aging of the nation's population. That aging, however, is only partly due to falling mortality and mostly to changes in fertility rates: fertility was generally high when today's elderly were born and is low now, so children have decreased and the elderly have increased as percentages of the population. Death rates also fall as income rises, partly because wealthier people are best able to afford good health care and other death controls, but also because high income, high levels of education, and high occupational status combine to depress death rates. To a considerable degree, one's probability of dying is related to one's economic standing.

Mortality levels are unevenly distributed geographically, partly because of differences in racial and ethnic composition, in health care from locale to locale, in rural and urban and metropolitan and nonmetropolitan residence, and in other factors. Large parts of the South have especially high death rates, but so do many other places with significant inner-city poverty, high unemployment rates, pronounced family disorganization, and other problems. The lowest death rates are generally in the Midwest and the Pacific Coast states and other parts of the western third of the nation. Most differences, however, result from the quality of health and medical care at the local level, and places where improvements are most needed are often least likely to get them because of financial constraints.

Mortality trends show significant variations since 1900, especially a fairly consistent decline until the mid-1950s, very little change from then until the late 1960s, a new drop after 1968 as various federal health programs (e.g., Medicare) took effect, and a tendency toward leveling and even some increases in mortality in the 1980s and early 1990s. The death rates of females fell more than those of males and the rates of blacks fell somewhat more than those of whites, so the sex differential increased and the racial differential decreased. Mortality decreases also fluctuated by age, but infants enjoyed larger reductions than most other age groups, the elderly the smallest reductions; all age groups, however, benefited significantly.

INFANT MORTALITY

Infant mortality is a sensitive indicator of the well-being of a population. It shows the United States to rank behind nearly two dozen other developed nations, most of which have significantly lower per capita GNP but more evenly accessible prenatal and other health care for mothers and infants.

The most dangerous time for newborns is immediately after birth, for well over half the deaths of infants aged 0–1 occur in the first six days, though at all ages in the first year the toll is greater among male than female infants. As with the loss of fetuses late in pregnancy, early infant mortality is high when prenatal care is poor, infants are born prematurely, are significantly underweight, and are born to unmarried mothers. As infants grow toward their first birthday, their chances of survival improve steadily, though variously affected by these conditions, along with the mother's age, her prior history of fetal or infant deaths, her lifestyle, and the father's genetic contribution. Since these conditions are less favorable on the average for blacks than whites, the black infant mortality rate is about twice that of whites and is much higher than for any other race or ethnic group, including Hispanics and Native Americans. Income and poverty play a part, as do levels of education, unemployment rates and their consequences for health care, and especially family structure. The large increase in the proportion of families headed by a woman without a husband present has had an especially pernicious effect on the survival prospects of infants, largely because so many negative factors coalesce in such families. Since the female-headed family is far more common among blacks than whites, particularly among the inner-city poor, it too contributes to the high infant mortality rate among blacks.

Wide geographic variations in infant mortality resemble those in general mortality and for many of the same reasons, although infant mortality rates are far more similar within each racial group than between them, regardless of whether the geographic unit is the region, division, state, or county. Nevertheless, geographic differences also reflect substantial local variations in prenatal care and, therefore, in the incidence of immaturity, low birth weight, and other conditions that affect infant mortality. At present, care is left largely to states and local areas, so the lack of uniform national guidelines and inadequate funding are reflected in the infant mortality mosaic, even though the overall high infant mortality rate is generally lamented as a national problem.

Causes of infant mortality fall into four large categories: congenital anomalies, sudden infant death syndrome (SIDS), certain conditions of the perinatal period, and preventable causes (e.g., infectious diseases, accidents, and homicide). Causes vary substantially between the neonatal period (first 27 days after birth) and the postneonatal period (28 days to the end of the first year). Neonatals are most likely to succumb to "endogenous" causes: congenital anomalies, problems relating to immaturity and low birth weight, respiratory distress syndrome, maternal complications of pregnancy, and complications of the placenta, umbilical cord, and membranes. Postneonatals are most apt to die of "exogenous" causes: SIDS, accidents, pneumonia, septicemia, and homicide, though many succumb to the lingering effects of congenital anomalies.

The infant mortality rate is only about a tenth that of early in the century, but neonatals now receive an inordinate amount of medical attention, especially to save severely underweight and immature infants, and postneonatals need more attention. In the 1980s overall progress in lowering infant mortality slowed and the high rate among blacks fell less than the much lower one among whites, so the racial gap widened. The incidence of low birth weight and immaturity did not change much, which contributed to the slow progress among blacks. Regardless of race, however, infant mortality among affluent Americans may be close to an irreducible minimum, whereas among the poor, especially teenagers, the rate remains excessive.

CAUSE OF DEATH

Cause-of-death analyses, using aggregate data, are complicated by attributing deaths to a single underlying cause, whereas individual deaths often result from multiple causes, especially among the elderly. Age-adjusted death rates for underlying causes, however, produce the following ranking for all Americans collectively: (1) diseases of the heart; (2) malignant neoplasms (cancer); (3) accidents and adverse effects; (4) cerebrovascular diseases (stroke); (5) chronic obstructive pulmonary diseases, such as emphysema, asthma, and bronchitis; (6) pneumonia and influenza; (7) suicide; (8) diabetes mellitus; (9) chronic liver disease and cirrhosis; and (10) homicide and legal intervention, although homicide is 99.9 percent of the category. Other significant causes include HIV/AIDS, drug- and alcohol-induced causes, and Alzheimer's disease. The death rate is increasing faster for HIV/AIDS than for any other cause, especially in population segments least affected previously, such as women, heterosexual teenagers, and infants.

Apart from sex-specific causes (e.g., prostate cancer, cervical cancer), the age-adjusted death rate for each cause is higher among males than females. The rate for most causes is also significantly higher among blacks than whites; blacks are six times more likely to die from homicide and twice as likely to die from diabetes and stroke. Of the ten major causes, the death rate is lower among blacks only for obstructive pulmonary diseases and suicide. The impact of various diseases also varies widely by age, regardless of race. Death rates for most causes are higher among infants than older children, who are most likely to die from accidents. After age 14 the death rates for several causes begin to rise, but accidents, especially by motor vehicle, continue to lead the list among persons aged 15–24, followed by homicide and suicide—all preventable causes that reflect the level of violence and recklessness in that age group. The ranking remains the same in the ages 25–34 and HIV/AIDS moves up in the list, especially for black and Hispanic males, but also for non-Hispanic white males. The degenerative diseases then take an increasing toll with age, and cancer heads the list among persons aged 35–64; lung cancer is the leading malignancy and its incidence is increasing among black and white women. It, combined with breast cancer and certain other malignancies, pushes the overall cancer death rate among white women aged 30–49 above that among white men, which is a rare reversal of the usual sex differential in mortality for nearly all causes. The same happens for black women aged 25–39. Heart disease attains first place as a cause of death among males 40–44 and among females 70–74, when it finally surpasses cancer. Thereafter, heart disease remains first, and since deaths are heavily concentrated in the oldest ages, heart disease is the nation's major cause of death according to the age-adjusted death rate and some age-specific rates.

TRENDS IN CAUSE OF DEATH

Long-term trends in cause of death reflect the nation's movement through the epidemiologic transition, from the early 1900s when infectious and parasitic illnesses claimed many lives, to the present when the degenerative diseases take a heavy toll. The HIV/AIDS epidemic, however, shows that we can never assume permanent immunity from new or resurgent communicable diseases and that only vigilant prevention, including childhood immunization against several diseases, can avoid new threats from them.

Trends in cardiovascular diseases show substantial death rate reductions for stroke and heart disease, including ischemic heart disease, heart failure, and hypertensive heart disease. Reductions in the incidence of smoking, obesity, high serum cholesterol, and high blood pressure helped lower death rates for heart disease and stroke, but the lower death rates also reflect the extent to which health-care expenditures have been focused on the elderly, in some respects at the expense of children. Despite the focus, the reported death rate for heart disease rose in the ages 85 and over, but while there were some real increases in that group, especially among blacks and women living alone, other apparent increases reflect faulty data. In addition, more persons in that age group are 90, 95, and even 100, and their death rates are high enough to increase the rate for the entire category. This is another consequence of the aging of the population.

Trends in death rates for cancer are less encouraging than for heart disease and stroke. The age-adjusted cancer death rate began to increase for some groups several decades ago, and in the 1980s it rose for black and white males and females, though by different degrees. This increase was the result of reductions in death rates at certain ages for certain types of cancer and increases at other ages for other types, most notably lung cancer as a consequence of smoking and other respiratory pollutants. The increase was greatest among middle-aged and elderly persons, especially women, because of the long-term carcinogenic effect of cigarette smoke. Furthermore, the death rate for female breast cancer failed to decrease among whites and increased among blacks, although this pattern also varies by age. The changes in other forms of cancer—colorectal, genitourinary, leukemia, and others—represent progress and regress, depending on age, race, sex, and other characteristics, but overall the decreases could not offset the increases and cancer is proportionately a more serious killer than it was.

Long-term trends in death rates for diabetes show significant reductions in most groups, though a large increase among black males; short-term trends show increases among other groups that keep the disease among the top ten killers. Death rates for chronic obstructive pulmonary diseases also rose, especially among the elderly because of their earlier smoking behavior. Death rates for pneumonia and liver diseases, including cirrhosis, fell, as did those for atherosclerosis and nephritis and related kidney diseases, though not for all groups or at all ages. Overall, changes in these causes of death also represent progress and regress.

Death rates for accidents, suicide, and homicide, all of which represent risks in the sociocultural environment, changed differently. Rates for accidents decreased among black and white males and females, though less for motor vehicle accidents than other types. Over the long term, suicide rates fell among whites of both sexes but rose among black males and females, reflecting significant changes in the relative socioeconomic and other conditions of the races. In the 1980s, however, suicide rates rose for several age groups of white males, including boys 5–14 and men from middle through old age, but they fell for white females of all ages. Other variations occurred among blacks that reflect particularly vulnerable ages. The most spectacular increases in the suicide rate are among children and adolescents, but the numbers are still small. The elderly recently reversed a 50-year decline, and their suicide rates began to rise. Homicide rates rose for all race and sex groups, though with some decreases in the 1980s among persons 35–84. Among young people and the very old, however, homicide rates went up because of several factors that affect various groups and persons differently. They include levels of self-esteem, the threshold for violence, structural poverty, divorce, and rural or urban residence, each of which is enmeshed in a highly complex network consisting of the other factors.

Death rates for most causes fell faster among women than men and the male/female ratio rose; rates for many causes fell faster among whites than blacks and the black/white mortality ratio rose, even though the racial ratio for all causes together fell somewhat. There are significant exceptions to these generalizations, however, and in the case of race the improvements for black males were less impressive than for black females or whites of either sex.

LIFE EXPECTANCY

While life expectancy decreases for each succeeding age group, the pace of decrease quickens steadily with age, thus reflecting rising death rates. Blacks have significantly lower life expectancy than whites at every age, until the crossover in the oldest years, and males have lower life expectancy than females at every age, though the gap diminishes with age. This reflects a broader tendency toward homogenization in health, longevity, morbidity, and mortality among the elderly because many biological factors grow more similar by sex and environmental factors pose somewhat more similar risks for elderly whites and blacks of both sexes. Thus, many sharp differences in life expectancy among younger persons decrease significantly among older ones. Differential survivorship, however, does produce a large "surplus" of widows, with implications for remarriage, loneliness, economic deprivation, and other realities that affect many elderly women living alone.

Life expectancy differs geographically for the same reasons that mortality differs, and the southern states generally rank relatively low in longevity, partly because of lower life expectancy in their black populations. That factor plays a significant role elsewhere too, and is especially pronounced in the nation's capital.

Life expectancy is closely related to cause of death, and by hypothetically eliminating specific causes we can estimate the years by which life expectancy would be raised. Four factors influence the increase: (1) If the death rate for a cause is quite high, eliminating it raises life expectancy most, though even eliminating heart disease adds only a few years to life expectancy. (2) If a large percentage of all deaths results from a cause, eliminating it adds most to life expectancy, though the additions are modest even for heart disease, cancer, and stroke, and minor for most other causes. (3) If a population experiencing the savings has overall high mortality, relatively little is added to life by eliminating one cause because persons soon die from something else. Conversely, if overall mortality is low, eliminating the most virulent causes adds more to life expectancy. (4) If the average age of death for a cause is low, eliminating it adds significantly to life expectancy. Eliminating conditions of the perinatal period, for example, leaves infants with virtually the entire average life expectancy before them, whereas eliminating heart disease among the elderly does not increase their survivorship much—they soon die of cancer, stroke, or other causes.

Those under age 65 presumably are still an economically productive group, so deaths among them constitute "premature mortality" and actual reductions in their mortality do add years of potentially productive life. For that reason, society is well served by aiming many lifesaving measures at the young, even though in the United States the elderly receive a disproportionately large share of the medical care and other available health resources and funds. Therefore, we do well to focus on increasing not just life expectancy, but "active life expectancy" and "years of functional health." We should be cautious about extending life expectancy in ways that raise the incidence of dependency, illness, disability, and despondency, especially at the cost of universal care for infants, children, and productive adults. Many life-sustaining technologies are applied largely to the very old to win a few additional weeks or months of life, and we need to decide whether the result is worth the cost, although many subjective considerations inevitably affect the decision.

Life expectancy is closely related to various health goals articulated by the Surgeon General in 1979 for 1990. Most of these goals, which relate to levels of health and mortality, were met for whites but not blacks. In 1990 the Secretary of Health and Human Services set new goals for 2000, largely oriented to improving preventive health measures and reducing health and mortality differentials among various groups. His overall intent was to increase active life expectancy at all ages and to emphasize effective prevention as a way to contain high treatment costs.

Long-term trends in life expectancy show substantial increases for all groups, though the pace of increase fluctuated in ways that parallel mortality changes. Infectious and parasitic illnesses were greatly reduced and life expectancy rose accordingly. The major increases occurred among infants, not the elderly, and life expectancy at birth rose more because a larger share of newborns survived to adulthood than because a large share of the elderly lived many additional years, though both factors played a part. Trends since 1970 show that white females continue to enjoy far higher life expectancy than any other groups; black females came to have higher life expectancy than white males; blacks of both sexes experienced some reductions in life expectancy after 1985; young black males fared the worst. The 1980s were not particularly kind to many groups, however, especially compared with improvements at other times during the century, and socioeconomic class divisions are clearly reflected in recent longevity changes. Census Bureau projections to 2080 assume that increases in life expectancy will be much more modest than between 1900 and 1990, that racial differences in longevity will disappear, and that the sex differential will persist. Short of a major social revolution, however, the latter projection seems more likely than the former.

Bibliography

Adams, Robert. "Address to the American Heart Association Writers' Seminar." Savannah, GA: January 16, 1991.

Alderson, Michael. *Mortality, Morbidity and Health Statistics*. New York: Stockton Press, 1988.

Allen, David M., James W. Buehler, Carol J. R. Hogue, Lilo T. Strauss, and Jack C. Smith. "Regional Differences in Birth Weight–Specific Infant Mortality, United States, 1980." *Public Health Reports* 102:2 (March-April, 1987): 138–145.

Anderson, John E., Laura Kann, Deborah Holtzman, Susan Arday, Ben Truman, and Lloyd Kolbe. "HIV/AIDS Knowledge and Sexual Behavior Among High School Students." *Family Planning Perspectives* 22:6 (November/December 1990): 252–255.

Anderson, Keavan M., William P. Castelli, and Daniel Levy. "Cholesterol and Mortality: Thirty Years of Follow-Up from the Framingham Study." *Journal of the American Medical Association* 257:16 (April 24, 1987): 2176–2180.

Archer, Dane, and Rosemary Gartner. *Violence and Crime in Cross-National Perspective*. New Haven, CT: Yale University Press, 1984.

Arriaga, Eduardo E. "Measuring and Explaining the Change in Life Expectancies." *Demography* 21:1 (February 1984): 83–96.

Assistant Secretary for Health and Surgeon General. *Healthy People: The Surgeon General's Report on Health Promotion and Disease Prevention*. Washington, DC: Government Printing Office, 1979.

Bachman, Jerald G., and Lloyd D. Johnston. "Patterns of Drug Use." *Institute for Social Research Newsletter* 15:2/3 (Fall/Winter 1987–88): 3, 6.

Baldwin, Wendy H., and Christine Winquist Nord. "Delayed Childbearing in the U.S.: Facts and Fictions." *Population Bulletin* 39:4. Washington, DC: Population Reference Bureau, 1985.

Bateson, Mary Catherine, and Richard Goldsby. *Thinking A.I.D.S.* Reading, MA: Addison-Wesley, 1988.

Berkman, Lisa, Burton Singer, and Kenneth Manton. "Black/White Differences in Health Status and Mortality Among the Elderly." *Demography* 26:4 (November 1989): 661–678.

Berkow, Robert, ed. *The Merck Manual.* 14th ed. Rahway, NJ: Merck Sharp & Dohme Research Laboratories, 1982.

Blakeslee, Sandra. "Father Figures: The Male Link to Birth Defects." *American Health* 10:3 (April 1991): 54–57.

Boone, Margaret S. *Capital Crime: Black Infant Mortality in America.* Newbury Park, CA: Sage, 1989.

Boserup, Ester. *Population and Technological Change: A Study of Long-Term Trends.* Chicago: University of Chicago Press, 1981.

Boulier, Bryan L., and Vincente B. Paqueo. "On the Theory and Measurement of the Determinants of Mortality." *Demography* 25:2 (May 1988): 249–263.

Brown, Lester R., and Jodi L. Jacobson. *Our Demographically Divided World.* Worldwatch Paper 74. Washington, DC: Worldwatch Institute, 1986.

Bruno, Kristen. "AIDS Update." *American Health* 10:3 (April 1991): 9.

Buehler, James W., Owen J. Devine, Ruth L. Berkelman, and Frances M. Chevarley. "Impact of the Human Immunodeficiency Virus Epidemic on Mortality Trends in Young Men, United States." *American Journal of Public Health* 80:9 (September 1990): 1080–1086.

Castelli, William P. *The Fundamental Role of Serum Lipids in Coronary Heart Disease.* New York: Pfizer, 1984.

Chambers, Robert. *Rural Development: Putting the Last First.* White Plains, NY: Longman, 1983.

Coale, Ansley J., and Graziella Caselli. "Estimation of the Number of Persons at Advanced Ages from the Number of Deaths at Each Age in the Given and Adjacent Years." *Genus* 46:1 (January-June 1990): 1–23.

———, and Guang Guo. "Revised Regional Model Life Tables at Very Low Levels of Mortality." *Population Index* 55:4 (Winter 1989): 613–643.

———, and Ellen Eliason Kisker. "Mortality Crossovers: Reality or Bad Data?" *Population Studies* 40:3 (November 1986): 389–401.

Cramer, James C. "Social Factors and Infant Mortality: Identifying High-Risk Groups and Proximate Causes." *Demography* 24:3 (August 1987): 299–322.

Crane, Jonathan. "The Epidemic Theory of Ghettos and Neighborhood Effects on Dropping Out and Teenage Childbearing." *American Journal of Sociology* 96:5 (March 1991): 1226–1259.

Crimmins, Eileen M., Yasuhiko Saito, and Dominique Ingegneri. "Changes in Life Expectancy and Disability-Free Life Expectancy in the United States." *Population and Development Review* 15:2 (June 1989): 235–267.

Daly, Martin, and Margo Wilson. *Homicide.* Hawthorne, NY: Aldine de Gruyter, 1988.

Datan, Nancy, and Nancy Lohman, eds. *Transitions of Aging.* New York: Academic Press, 1980.

Davis, Cary, Carl Haub, and JoAnne Willette. "U.S. Hispanics: Changing the Face of America." *Population Bulletin* 38:3. Washington, DC: Population Reference Bureau, 1983.

Dublin, Louis J., Alfred J. Lotka, and Mortimer Spiegelman. *Length of Life.* New York: Ronald Press, 1949.

Duchene, Josianne, and Guillaume Wunsch. "From the Demographer's Cauldron: Single

Decrement Life Tables and the Span of Life.'' *Genus* 44:3 (July–December 1985): 1–17.

Duleep, Harriet Orcutt. ''Measuring Socioeconomic Mortality Differentials Over Time.'' *Demography* 26:2 (May 1989): 345–351.

Duncan, Greg J., and Saul D. Hoffman. ''A Reconsideration of the Economic Consequences of Marital Dissolution.'' *Demography* 22:4 (November 1987): 485–497.

———. ''Welfare Benefits, Economic Opportunities, and Out-of-Wedlock Births Among Black Teenage Girls.'' *Demography* 27:4 (November 1990): 519–535.

Durkheim, Emile. *Suicide*, translated by John A. Spaulding and George Simpson. New York: Free Press, 1951.

Eberstein, Isaac W. ''Demographic Research on Infant Mortality.'' *Sociological Forum* 4:3 (September 1989): 409–422.

———, Charles B. Nam, and Robert A. Hummer. ''Infant Mortality by Cause of Death: Main and Interaction Effects.'' *Demography* 27:3 (August 1990): 423–430.

Friede, Andrew, Wendy Baldwin, Philip H. Rhodes, James W. Buehler, Lilo T. Strauss, Jack C. Smith, and Carol J. R. Hogue. ''Young Maternal Age and Infant Mortality: The Role of Low Birth Weight.'' *Public Health Reports* 102:2 (March-April 1987): 192–199.

Gabennesch, Howard. ''When Promises Fail: A Theory of Temporal Fluctuations in Suicide.'' *Social Forces* 67:1 (September 1988): 129–145.

Gartner, Rosemary. ''The Victims of Homicide: A Temporal Cross-National Comparison.'' *American Sociological Review* 55:1 (February 1990): 92–106.

———, and Robert Nash Parker. ''Cross-National Evidence on Homicide and the Age Structure of the Population.'' *Social Forces* 69:2 (December 1990): 351–371.

Geronimus, Arline T. ''On Teenage Childbearing and Neonatal Mortality in the United States.'' *Population and Development Review* 13:2 (June 1987): 245–279.

———, and John Bound. ''Black/White Differences in Women's Reproductive Health Status: Evidence From Vital Statistics.'' *Demography* 27:3 (August 1990): 457–466.

Gibbs, Jack P., ed. *Suicide*. New York: Harper & Row, 1968.

Goodwin, Paul B., Jr., ed. *Latin America*. 3d ed. Guilford, CT: Dushkin Publishing Group, 1988.

Gordon, Theodore J. ''Prospects for Aging in America.'' In *Aging from Birth to Death*, edited by Matilda White Riley. Boulder, CO: Westview Press, 1979.

Gould, Madelyn S., Sylvan Wallenstein, Marjorie H. Kleinman, Patrick O'Carroll, and James Mercy. ''Suicide Clusters: An Examination of Age-Specific Effects.'' *American Journal of Public Health* 80:2 (February 1990): 211–212.

Gove, Walter R. ''Sex, Marital Status, and Mortality.'' *American Journal of Sociology* 79:1 (July 1973): 45–67.

Grove, Robert D., and Alice M. Hetzel. *Vital Statistics Rates in the United States, 1940–1960*. Washington, DC: Government Printing Office, 1968.

Hale, Christiane B. *Infant Mortality: An American Tragedy*. Occasional paper 13 in the series Population Trends and Public Policy. Washington, DC: Population Reference Bureau, 1988.

Hall, Jeff M., Ming K. Lee, Beth Newman, Jan E. Morrow, Lee A. Anderson, Bing Huey, and Mary-Claire King. ''Linkage of Early-Onset Familial Breast Cancer to Chromosome 17q21.'' *Science* 250:4988 (December 21, 1990): 1684–1689.

Hansotia, Phiroze, and Steven K. Broste. ''The Effect of Epilepsy or Diabetes Mellitus on the Risk of Automobile Accidents." *New England Journal of Medicine* 324:1 (January 3, 1991): 22–26.

Hasenfeld, Yeheskel, and Mayer N. Zald, eds. "The Welfare State in America: Trends and Prospects." *Annals of the American Academy of Political and Social Science* 479 (May 1985): 1–206.

Haub, Carl, and Machiko Yanagashita. "Infant Mortality: Who's Number One?" *Population Today* 19:30 (March 1991): 6–8.

Hayflick, Leonard. "Human Cells and Aging." *Scientific American* 218:3 (March 1968): 32–37.

Henderson, Brian, Annlia Paganini-Hill, and Ronald K. Ross. "Decreased Mortality in Users of Estrogen Replacement Therapy." *Archives of Internal Medicine* 151:1 (January 4, 1991): 75–78.

Henry, Louis. "Men's and Women's Mortality in the Past." *Population* 44:1 (September 1989): 177–201.

Hickey, Tom, William Rakowski, and Mara Julius. "Preventive Health Practices Among Older Men and Women." *Research on Aging* 10:3 (September 1988): 315–328.

Higgins, Millicent W., and Russell V. Luepker. *Trends in Coronary Heart Disease Mortality: The Influence of Medical Care.* New York: Oxford University Press, 1988.

Hill, Kenneth, and Anne R. Pebley. "Child Mortality in the Developing World." *Population and Development Review* 15:4 (December 1989): 657–687.

Hill, Martha S. "The Changing Nature of Poverty." In "The Welfare State in America: Trends and Prospects," edited by Yeheskel Hasenfeld and Mayer N. Zald. *Annals of the American Academy of Political and Social Science* 479 (May 1985): 31–47.

Hogue, Carol J. R., James W. Buehler, Lilo T. Strauss, and Jack C. Smith. "Overview of the National Infant Mortality Surveillance (NIMS) Project—Design, Methods, Results." *Public Health Reports* 102:2 (March-April 1987): 126–138.

Hollingsworth, J. Selwyn, Everett S. Lee, Douglas C. Bachtel, Mohamed El-Attar, and C. Jack Tucker. "Infant Mortality in the Deep South." Papers presented at the Southern Demographic Association. Durham, NC: October 18–20, 1989.

Horiuchi, Shiro, and Ansley J. Coale. "A Simple Equation for Estimating the Expectation of Life at Old Ages." *Population Studies* 36:2 (July 1982): 317–326.

Hu, Yuanreng, and Noreen Goldman. "Mortality Differentials by Marital Status: An International Comparison." *Demography* 27:2 (May 1990): 233–250.

Institute of Medicine. *Preventing Low Birth Weight.* Washington, DC: National Academy of Sciences Press, 1986.

International Planned Parenthood Federation. Tony Klouda. " 'Marginalization,—Useful Concept for AIDS Projects." *AIDS Watch* 13 (1991): 2.

Jackson, Jacquelyn Johnson. *Minorities and Aging.* Belmont, CA: Wadsworth, 1980.

Janerich, Dwight T., W. Douglas Thompson, Luis R. Varela, Peter Greenwald, Sherry Chorost, Cathy Tucci, Muhammed B. Zaman, Myron E. Melamed, Maureen Kiely, and Martin F. McKneally. "Lung Cancer and Exposure to Tobacco Smoke in the Household." *New England Journal of Medicine* 323:10 (September 6, 1990): 632–636.

Jensen, Leif, and Marta Tienda. "Nonmetropolitan Families in the United States: Trends in Racial and Ethnic Stratification." *Rural Sociology* 54:4 (Winter 1989): 509–532.

Katz, Sidney, Laurence G. Branch, Michael H. Branson, Joseph H. Papsidero, John C. Beck, and David S. Greer. "Active Life Expectancy." *New England Journal of Medicine.* 309:20 (November 17, 1983): 1218–1224.

Keith, Verna M., and David P. Smith. "The Current Differential in Black and White Life Expectancy." *Demography* 25:4 (November 1988): 625–632.

Keyfitz, Nathan. "What Difference Would It Make If Cancer Were Eradicated? An Examination of the Taeuber Paradox." *Demography* 14:4 (November 1977): 411–418.

Kitagawa, Evelyn M., and Philip M. Hauser. *Differential Mortality in the United States: A Study in Socioeconimic Epidemiology.* Cambridge, MA: Harvard University Press, 1973.

Kleinman, Joel C. "Infant Mortality Among Racial/Ethnic Minority Groups, 1983–1984." *Morbidity and Mortality Weekly Report* 39:553 (July 1990): 31–39.

Kohn, Robert R. "Cause of Death in Very Old People." *Journal of the American Medical Association* 247:20 (May 28, 1982): 2793–2797.

Kowalski, Gregory S., and Don Duffield. "The Impact of the Rural Population Component on Homicide Rates in the United States: A County-Level Analysis." *Rural Sociology* 55:1 (Spring 1990): 76–90.

Landesman, Sheldon H., Harold M. Ginzberg, and Stanley H. Weiss. "The AIDS Epidemic." *New England Journal of Medicine* 312:8 (February 21, 1985): 521–525.

Lopata, Helen Znaniecki. "The Widowed Family Member." In *Transitions of Aging,* edited by Nancy Datan and Nancy Lohman. New York: Academic Press, 1980.

Lopez, Alan D. "Who Dies of What? A Comparative Analysis of Mortality Conditions in Developed Countries Around 1987." *World Health Statistical Quarterly* 43:2 (1990): 104–114.

Madigan, Francis C. "Are Sex Mortality Differentials Biologically Caused?" *Milbank Memorial Fund Quarterly* 35:2 (April 1957): 202–223.

Manton, Kenneth G., and George C. Myers. "Recent Trends in Multiple-Caused Mortality 1968 to 1982: Age and Cohort Components." *Population Research and Policy Review* 6:2 (1987): 161–176.

———, and Eric Stallard. "Methods for the Analysis of Mortality Risks Across Heterogeneous Small Populations: Examination of Space-Time Gradients in North Carolina Counties, 1970–75." *Demography* 18:2 (May 1981): 217–230.

———, and Eric Stallard. *Recent Trends in Mortality Analysis.* New York: Academic Press, 1984.

———, Eric Stallard, Max A. Woodbury, Wilson B. Riggan, John P. Creason, and Thomas J. Mason. "Statistically Adjusted Estimates of Geographic Mortality Profiles." *Journal of the National Cancer Institute* 78:5 (May 1987): 805–815.

Marks, James S., James W. Buehler, Lilo T. Strauss, Carol J. R. Hogue, and Jack C. Smith. "Variation in State-Specific Infant Mortality Risks." *Public Health Reports* 102:2 (April-March 1987): 146–151.

Martin, Barbara V., and Gerardo Marin. "Hispanics and AIDS." Special issue of *Hispanic Journal of Behavioral Sciences* 12:2 (May 1990).

Maulitz, Russell C., ed. *Unnatural Causes: The Three Leading Killer Diseases in America.* New Brunswick, NJ: Rutgers University Press, 1989.

McCormick, Marie C. "The Contribution of Low Birth Weight to Infant Mortality and Childhood Morbidity." *New England Journal of Medicine* 312:2 (January 2, 1985): 88–90.

McFarlan, Donald, ed. *The Guinness Book of Records, 1991.* New York: Facts on File, 1990.

McNamara, Regina. "Infant and Child Mortality." In *International Encyclopedia of Population.* Vol. 1. Edited by John A. Ross. New York: Free Press, 1982.

———. "Mortality Trends: Historical Trends." In *Encyclopedia of Population.* Vol. 2. Edited by John A. Ross. New York: Free Press, 1982.

Messner, Steven. "Economic Discrimination and Societal Homicide Rates: Further Evidence on the Cost of Inequality." *American Sociological Review* 54:4 (August 1989): 597–611.

Metropolitan Life Insurance Company. "Deaths from Chronic Obstructive Pulmonary Disease in the United States, 1987." *Statistical Bulletin* 71:3 (July-September 1990): 20–26.

———. "Women's Longevity Advantage Declines." *Statistical Bulletin* 69:1 (January-March 1988): 18–23.

Mitra, Samarendranath. "Estimating the Expectation of Life at Older Ages." *Population Studies* 38:2 (July 1984): 313–319.

Nam, Charles B., and Susan Gustavus Philliber. *Population: A Basic Orientation.* 2d ed. Englewood Cliffs, NJ: Prentice-Hall, 1984.

———, Norman L. Weatherby, and Kathleen A. Ockay. "Causes of Death Which Contribute to the Mortality Crossover Effect." *Social Biology* 25:4 (Winter 1978): 306–314.

National Center for Health Statistics. "Advance Report of Final Mortality Statistics, 1986." *Monthly Vital Statistics Report* 37:6. Washington, DC: Government Printing Office, September 1988.

———. "Advance Report of Final Mortality Statistics, 1987." *Monthly Vital Statistics Report* 38:5. Washington, DC: Government Printing Office, September 1989.

———. "Advance Report of Final Mortality Statistics, 1988." *Monthly Vital Statistics Report* 39:7. Washington, DC: Government Printing Office, November 1990.

———. "Advance Report of Final Natality Statistics, 1988." *Monthly Vital Statistics Report* 39:4. Washington, DC: Government Printing Office, August 1990.

———. "Annual Summary of Births, Marriages, Divorces, and Deaths: United States, 1989." *Monthly Vital Statistics Report* 38:13. Washington, DC: Government Printing Office, August 1990.

———. "Current Estimates from the National Health Interview Survey: United States, 1987." *Vital and Health Statistics,* Series 10, No. 166. Washington, DC: Government Printing Office, 1988.

———. *Health, United States, 1979.* Washington, DC: Government Printing Office, 1980.

———. *Health, United States, 1981.* Washington, DC: Government Printing Office, 1981.

———. *Health, United States, 1987.* Washington, DC: Government Printing Office, 1988.

———. *Health, United States, 1988.* Washington, DC: Government Printing Office, 1989.

———. *Health, United States, 1989.* Washington, DC: Government Printing Office, 1990.

———. *Health, United States, 1990.* Washington, DC: Government Printing Office, 1991.

————. "State Life Tables." Florida. *U.S. Decennial Life Tables for 1979–81.* Vol. II, No. 10. Washington, DC: Government Printing Office, 1985.

————. *U.S. Decennial Life Tables for 1979–81.* Vol. I, No. 4. Washington, DC: Government Printing Office, 1988.

————. *Vital Statistics of the United States, 1979.* Vol. II. *Mortality.* Part A. Washington, DC: Government Printing Office, 1984.

————. *Vital Statistics of the United States, 1985.* Vol. I. *Natality.* Washington, DC: Government Printing Office, 1988.

————. *Vital Statistics of the United States, 1985.* Vol. II. *Mortality.* Parts A, B. Washington, DC: Government Printing Office, 1988.

————. *Vital Statistics of the United States, 1986.* Vol. I. *Natality.* Washington, DC: Government Printing Office, 1988.

————. *Vital Statistics of the United States, 1986.* Vol. II. *Mortality.* Parts A, B. Washington, DC: Government Printing Office, 1988.

————. *Vital Statistics of the United States, 1987.* Vol. I. *Natality.* Washington, DC: Government Printing Office, 1989.

————. *Vital Statistics of the United States, 1987.* Vol. II. *Mortality.* Parts A, B. Washington, DC: Government Printing Office, 1990.

————. *Vital Statistics of the United States, 1988.* Vol. II. *Mortality.* Part B. Washington, DC: Government Printing Office, 1991.

————. Robert J. Armstrong. "Some Trends and Comparisons of United States Life Table Data: 1900–1981." *U.S. Decennial Life Tables for 1979–81.* Vol. I, No. 4. Washington, DC: Government Printing Office, 1987.

————. Barbara Bloom. "Health Insurance and Medical Care: Health of Our Nation's Children, United States, 1988." *Advance Data from Vital and Health Statistics* 188. Washington, DC: Government Printing Office, October 1990.

————. Lester R. Curtin and Robert J. Armstrong. "United States Life Tables Eliminating Certain Causes of Death." *U.S. Decennial Life Tables for 1979–81.* Vol. 1, No. 2. Washington, DC: Government Printing Office, 1988.

————. Beulah K. Cypress. "Health Care of Adolescents by Office-Based Physicians: National Ambulatory Medical Care Survey, 1980–81." *Advance Data from Vital and Health Statistics* 99. Washington, DC: Government Printing Office, September 1984.

————. Deborah A. Dawson and Ann M. Hardy. "AIDS Knowledge and Attitudes of Black Americans." *Advance Data from Vital and Health Statistics* 165. Washington, DC: Government Printing Office, March 1989.

————. Deborah A. Dawson and Ann M. Hardy. "AIDS Knowledge and Attitudes of Hispanic Americans." *Advance Data from Vital and Health Statistics* 166. Washington, DC: Government Printing Office, April 1989.

————. Thomas F. Drury and Anita L. Powell. "Prevalence, Impact, and Demography of Known Diabetes in the United States." *Advance Data from Vital and Health Statistics* 114. Washington, DC: Government Printing Office, February 1986.

————. Thomas F. Drury and Anita L. Powell. "Prevalence of Known Diabetes Among Black Americans." *Advance Data from Vital and Health Statistics* 130. Washington, DC: Government Printing Office, July 1987.

————. Thomas F. Drury and Ildy I. Shannon. "Health Practices and Perceptions of U.S. Adults with Noninsulin Dependent Diabetes: Data from the 1985 National Health Interview Survey of Health Promotion and Disease Prevention." *Advance*

Data from Vital and Health Statistics 141. Washington, DC: Government Printing Office, September 1987.

———. Lois Fingerhut. "Trends and Current Status in Childhood Mortality, United States, 1980–85." *Vital and Health Statistics*, Series 3, No. 26. Washington, DC: Government Printing Office, 1989.

———. Joseph E. Fitti and Marcie Cynamon. "AIDS Knowledge and Attitudes for April-June 1990." *Advance Data from Vital and Health Statistics* 195. Washington, DC: Government Printing Office, December 1990.

———. Jack M. Guralnick, Andrea Z. LaCroix, Donald F. Everett, and Mary Grace Kovar. "Aging in the Eighties: The Prevalence of Comorbidity and Its Association With Disability." *Advance Data from Vital and Health Statistics* 170. Washington, DC: Government Printing Office, May 1989.

———. Gloria Kapantais and Eve Powell-Griner. "Characteristics of Persons Dying from AIDS." *Advance Data from Vital and Health Statistics* 173. Washington, DC: Government Printing Office, August 1989.

———. Gloria Kapantais and Eve Powell-Griner. "Characteristics of Persons Dying of Diseases of Heart." *Advance Data from Vital and Health Statistics* 172. Washington, DC: Government Printing Office, August 1989.

———. Mary Grace Kovar. "Aging in the Eighties, Age 65 and Over and Living Alone, Contacts with Family, Friends, and Neighbors." *Advance Data from Vital and Health Statistics* 116. Washington, DC: Government Printing Office, May 1986.

———. Mary Grace Kovar. "Aging in the Eighties, People Living Alone—Two Years Later." *Advance Data from Vital and Health Statistics* 149. Washington, DC: Government Printing Office, April 1988.

———. Linda W. Pickle, Thomas J. Mason, Neil Howard, Robert Hoover, and Joseph S. Fraumeni. *Atlas of U.S. Cancer Mortality among Whites: 1950–1980.* Washington, DC: Government Printing Office, 1987.

———. Eve Powell-Griner. "Characteristics of Persons Dying from Cerebrovascular Diseases." *Advance Data from Vital and Health Statistics* 180. Washington, DC: Government Printing Office, February 1990.

———. William F. Pratt and Marjorie C. Horn. "Wanted and Unwanted Childbearing: United States, 1973–82." *Advance Data from Vital and Health Statistics* 108. Washington, DC: Government Printing Office, May 1985.

———. Peter Ries. "Health Care Coverage by Age, Sex, Race, and Family Income: United States, 1986." *Advance Data from Vital and Health Statistics* 139. Washington, DC: Government Printing Office, September 1987.

———. Charlotte A. Schoenborn and Bernice H. Cohen. "Trends in Smoking, Alcohol Consumption, and Other Health Practices Among U.S. Adults, 1977 and 1983." *Advance Data from Vital and Health Statistics* 118. Washington, DC: Government Printing Office, June 1986.

———. Owen T. Thornberry. "Health Promotion Data for the 1990 Objectives." *Advance Data from Vital and Health Statistics* 126. Washington, DC: Government Printing Office, September 1986.

Navarro, Vincente. "Race or Class Versus Race and Class: Mortality Differentials in the United States." *Lancet* 336:8725 (November 17, 1990): 1238–1240.

North Carolina Department of Environment, Health, and Natural Resources. Standard Certificate of Death, 1989 Revision. Raleigh, NC.

Novello, Antonia C., Paul H. Wise, and D. V. Kleinman. "Hispanic Health: Time for

Data, Time for Action." *Journal of the American Medical Association* 265:2 (January 9, 1991): 253–255.

Omran, Abdel R. "Epidemiologic Transition Theory." In *International Encyclopedia of Population*. Vol. 1. Edited by John A. Ross. New York: Free Press, 1982.

―――. "Epidemiologic Transition: A Theory of the Epidemiology of Population Change." *Milbank Memorial Fund Quarterly* 4:1 (October 1971): 509–538.

―――. "Epidemiologic Transition: United States." In *International Encyclopedia of Population*. Vol. 1. Edited by John A. Ross. New York: Free Press, 1982.

Parker, Robert Nash. "Poverty, Subculture of Violence, and Type of Homicide." *Social Forces* 67:4 (June 1989): 983–1007.

Paxton, John, ed. *The Statesman's Yearbook, 1988–89*. New York: St. Martin's Press, 1988.

―――. *The Statesman's Yearbook, 1990–91*. New York: St. Martin's Press, 1990.

Pekkanen, Juha, Shai Linn, Gerardo Heiss, Chirayath M. Suchindran, Arthur Leon, Basil F. Rifkind, and Herman Tryoler. "Ten-Year Mortality from Cardiovascular Disease in Relation to Cholesterol Level Among Men With and Without Preexisting Cardiovascular Disease." *New England Journal of Medicine* 322:24 (June 14, 1990): 1700–1717.

Pollard, A. H., Farhat Yusuf, and G. N. Pollard. *Demographic Techniques*. 3d ed. Sydney, Australia: Pergamon Press, 1990.

Pollard, John H. "On the Decomposition of Changes in Expectation of Life and Differentials in Life Expectancy." *Demography* 25:2 (May 1988): 265–276.

Population Reference Bureau. *World Population Data Sheet, 1988*. Washington, DC: Population Reference Bureau, 1988.

―――. *World Population Data Sheet, 1989*. Washington, DC: Population Reference Bureau, 1989.

―――. *World Population Data Sheet, 1990*. Washington, DC: Population Reference Bureau, 1990.

Potter, Lloyd B., and Omer R. Galle. "Residential and Racial Mortality Differentials in the South by Cause of Death." *Rural Sociology* 55:2 (Summer 1990): 233–244.

Powell-Griner, Eve, and Albert Woolbright. "Trends in Infant Deaths from Congenital Anomalies: Results from England and Wales, Scotland, Sweden, and the United States." *International Journal of Epidemiology* 19:2 (June 1990): 391–398.

Preston, Samuel H. *Mortality Patterns in National Populations: With Special Reference to Recorded Causes of Death*. New York: Academic Press, 1976.

―――, ed. *The Effects of Infant and Child Mortality on Fertility*. New York: Academic Press, 1978.

Ragland, Kathleen E., Steve Selvin, and Deane W. Merrill. "The Onset of Decline in Ischemic Heart Disease Mortality in the United States." *American Journal of Epidemiology* 127:3 (March 1988): 516–531.

Reich, Robert B. "What We Can Do." In *Crisis in American Institutions*. 7th ed. Edited by Jerome H. Skolnick and Elliott Currie. Glenview, IL: Scott, Foresman, 1988.

Reid, John. "Black America in the 1980s." *Population Bulletin* 37:4. Washington, DC: Population Reference Bureau, 1982.

Relman, Arnold S. "Is Rationing Inevitable?" *New England Journal of Medicine* 322:25 (June 21, 1990): 1809–1810.

Repetto, Robert. "Population, Resources, Environment: An Uncertain Future." *Population Bulletin* 42:2. Washington, DC: Population Reference Bureau, 1987.

Retherford, Robert D. *The Changing Sex Differential in Mortality.* Westport, CT: Greenwood Press, 1975.

———. "Tobacco Smoking and the Sex Mortality Differential." *Demography* 9:2 (May 1972): 203–216.

Richardson, Hal B. Critique of Juha Pekkanen, Shai Linn, Gerardo Heiss, Chirayath M. Suchindran, Arthur Leon, Basil F. Rifkind, and Herman Tryoler. "Ten-Year Mortality from Cardiovascular Disease in Relation to Cholesterol Level Among Men with and without Preexisting Cardiovascular Disease." *New England Journal of Medicine* 322:24 (June 14, 1990), pp. 1700–1717, in *New England Journal of Medicine* 324:1 (January 3, 1991): 60–61.

Riley, Matilda White, ed. *Aging from Birth to Death.* Boulder, CO: Westview Press, 1979.

———, Marcia G. Ory, and Diane Zablotsky, eds. *AIDS in an Aging Society: What We Need to Know.* New York: Springer, 1989.

Rosenwaike, Ira. "Introduction." In *Mortality of Hispanic Populations,* edited by Ira Rosenwaike. Westport, CT: Greenwood Press, 1991.

———. "Mortality Differentials among Persons Born in Cuba, Mexico, and Puerto Rico Residing in the United States." *American Journal of Public Health* 77:5 (May 1987): 603–606.

———, ed. *Mortality of Hispanic Populations.* Westport, CT: Greenwood Press, 1991.

———, and Katherine Hempstead. "Mortality Among Three Puerto Rican Populations: Residents of Puerto Rico and Migrants in New York City and the Balance of the United States, 1979–81." *International Migration Review* 24:4 (Winter 1990): 684–702.

Ross, John A. "Life Tables." In *International Encyclopedia of Population.* Vol. 2. Edited by John A. Ross. New York: Free Press, 1982.

———, ed. *International Encyclopedia of Population.* 2 Vols. New York: Free Press, 1982.

Rubin, Rita. "The Ways of a Woman's Heart." *American Health* 10:2 (March 1991): 6.

Saunders, John. *Basic Demographic Measures.* Lanham, MD: University Press of America, 1988.

Schneider, Edward L., and John D. Reed, Jr. "Life Extension." *New England Journal of Medicine* 312:18 (May 1, 1985): 1159–1168.

Schwartz, Eugene, Vincent Y. Kofie, Marc Rivo, and Reed V. Tucson. "Black/White Comparisons of Deaths Preventable by Medical Intervention: United States and the District of Columbia." *International Journal of Epidemiology* 19:3 (September 1990): 591–598.

Senderowitz, Judith, and John M. Paxman. "Adolescent Fertility: Worldwide Concerns." *Population Bulletin* 40:2. Washington, DC: Population Reference Bureau, 1985.

Shryock, Henry S., and Jacob S. Siegel. *The Methods and Materials of Demography.* 2 Vols. Washington, DC: Government Printing Office, 1973.

Skolnick, Jerome H., and Elliott Currie, eds. *Crisis in American Institutions.* 7th ed. Glenview, IL: Scott, Foresman, 1988.

Smil, Vaclav. "Coronary Heart Disease, Diet, and Western Mortality." *Population and Development Review* 15:3 (September 1989): 399–424.

Stamler, Jeremiah. "Coronary Heart Disease: Doing the Right Things." *New England Journal of Medicine* 312:16 (April 19, 1985): 1053–1055.

Stolnitz, George J. "A Century of International Mortality Trends: II." *Population Studies* 10:1 (March 1956): 17–42.

Streatfield, Kim, Masri Singarimbun, and Ian Diamond. "Maternal Education and Child Immunization." *Demography* 27:3 (August 1990): 447–455.

Sullivan, Louis. *Healthy People 2000: National Health Promotion and Disease Prevention Objectives*. Washington, DC: Government Printing Office, 1990.

———. "Healthy People 2000." *New England Journal of Medicine* 323:15 (October 11, 1990): 1065–1067.

Swanson, David A. "A State-Based Regression Model for Estimating Substate Life Expectancy." *Demography* 26:1 (February 1989): 161–170.

Thurmon, Theodore F., Susonne A. Ursin, and Kathleen R. Wiley. "Association of Lung Cancer Death Rates by Parish with Migration Rate by Age Group." *American Journal of Human Biology* 1:6 (1989): 771–784.

Turner, Charles F., Heather G. Miller, and Lincoln E. Moses, eds. *AIDS: Sexual Behavior and Intravenous Drug Use*. Washington, DC: National Academy Press, 1989.

United Nations. *Demographic Yearbook, 1988*. New York: United Nations, 1990.

U.S. Bureau of the Census. "Demographic Aspects of Aging and the Older Population in the United States." *Current Population Reports*, Series P–23, No. 59. Washington, DC: Government Printing Office, 1978.

———. "Educational Attainment in the United States: March 1987 and 1986." *Current Population Reports*, Series P–20, No. 428. Washington, DC: Government Printing Office, 1988.

———. "Hispanic Population of the United States: March 1989." *Current Population Reports*, Series P–20, No. 444. Washington, DC: Government Printing Office, 1990.

———. "Measuring the Effects of Benefits and Taxes on Income and Poverty: 1989." *Current Population Reports*, Series P–60, No. 169-RD. Washington, DC: Government Printing Office, 1990.

———. "Money Income of Households, Families, and Persons in the United States: 1987." *Current Population Reports*, Series P–60, No. 162. Washington, DC: Government Printing Office, 1989.

———. "Money Income and Poverty Status in the United States: 1988." *Current Population Reports*, Series P–60, No. 166. Washington, DC: Government Printing Office, 1989.

———. *1980 Census of Population. Race of the Population by States: 1980*. Supplementary Reports, PC-80–S1–3. Washington, DC: Government Printing Office, 1981.

———. "Population Profile of the United States: 1989." *Current Population Reports*, Series P–23, No. 159. Washington, DC: Government Printing Office, 1989.

———. "Poverty in the United States: 1987." *Current Population Reports*, Series P–60, No. 163. Washington, DC: Government Printing Office, 1989.

———. "Residents of Farms and Rural Areas: 1989." *Current Population Reports*, Series P–20, No. 446. Washington, DC: Government Printing Office, 1990.

———. *Sixteenth Census of the United States: 1940. Characteristics of the Population. United States Summary*. Washington, DC: Government Printing Office, 1943.

———. *Social Indicators III*. Washington, DC: Government Printing Office, 1980.

———. *Statistical Abstract of the United States: 1990*. Washington, DC: Government Printing Office, 1990.

———. "United States Population Estimates, by Age, Sex, Race, and Hispanic Origin: 1980 to 1988." *Current Population Reports,* Series P–25, No. 1045. Washington, DC: Government Printing Office, 1990.

———. Paul C. Glick. "The Future of the American Family." *Current Population Reports,* Series P–23, No. 78. Washington, DC: Government Printing Office, 1979.

———. Paul C. Glick. "Population of the United States, Trends and Prospects: 1950–1990." *Current Population Reports,* Series P–23, No. 49. Washington DC: Government Printing Office, 1974.

———. Ellen Jamison. *World Population Profile: 1989.* Washington, DC: Government Printing Office, 1989.

———. Ellen Jamison, Peter D. Johnson, and Richard A. Engels. *World Population Profile: 1987.* Washington, DC: Government Printing Office, 1987.

———. Arlene F. Saluter. "Marital Status and Living Arrangements: March 1979." *Current Population Reports,* Series P–20, No. 349. Washington, DC: Government Printing Office, 1980.

———. Arlene F. Saluter. "Marital Status and Living Arrangements: March 1988." *Current Population Reports,* Series P–20, No. 433. Washington, DC: Government Printing Office, 1989.

———. Arlene F. Saluter. "Marital Status and Living Arrangements: March 1989." *Current Population Reports,* Series P–20, No. 445. Washington, DC: Government Printing Office, 1990.

———. Gregory Spencer. "Projections of the Population of the United States, by Age, Sex, and Race: 1988 to 2080." *Current Population Reports,* Series P–25, No. 1018. Washington, DC: Government Printing Office, 1989.

U.S. Centers for Disease Control. "Smoking-Attributable Mortality and Potential Years of Life Lost—United States, 1984." *Morbidity and Mortality Weekly Report* 36:42 (October 1987): 693–697.

U.S. Congress. Office of Technology Assessment. *Life-Sustaining Technologies and the Elderly.* Washington, DC: Government Printing Office, 1987.

U.S. Department of Health and Human Services. *Report of the Secretary's Task Force on Black and Minority Health.* Washington, DC: Government Printing Office, 1985.

U.S. House of Representatives. Subcommittee on Human Services of the Select Committee on Aging. *Future Directions for Aging Policy: A Human Services Model.* Washington, DC: Government Printing Office, 1980.

Valdiserri, Ronald O. *Preventing AIDS: The Design of Effective Programs.* New Brunswick, NJ: Rutgers University Press, 1989.

Vaupel, James W. "Inherited Frailty and Longevity." *Demography* 25:2 (May 1988): 277–287.

———, and Anatoli I. Yashin. "Repeated Resuscitation: How Lifesaving Alters Life Tables." *Demography* 24:1 (February 1987): 123–135.

Verbrugge, Lois M. "A Health Profile of Older Women with Comparisons to Older Men." *Research on Aging* 6:3 (September 1984): 291–322.

Wallis, Claudia. "A Puzzling Plague." *Time* 137:2 (January 14, 1991): 48–52.

Weeks, John R. "The Demography of Islamic Nations." *Population Bulletin* 43:4. Washington, DC: Population Reference Bureau, 1988.

————. *Population: An Introduction to Concepts and Issues*. 4th ed. Belmont, CA: Wadsworth, 1989.

Weinberg, Robert A. "Oncogenes and Tumor Suppressor Genes." In *Unnatural Causes: The Three Leading Killer Diseases in America*, edited by Russell C. Maulitz. New Brunswick, NJ: Rutgers University Press, 1989.

Weinstein, Milton C., Pamela G. Coxson, Lawrence W. Williams, Theodore M. Pass, William B. Stason, and Lee Goldman. "Forecasting Coronary Heart Disease Incidence, Mortality, and Cost: The Coronary Heart Disease Model." *American Journal of Public Health* 77:11 (November 1987): 417–426.

Wilkinson, Kenneth P. "A Research Note on Homicide and Rurality." *Social Forces* 63:2 (December 1984): 445–452.

Willett, Walter C., Meir J. Stampfer, Graham A. Colditz, Bernard A. Rosner, and Frank E. Speizer. "Relation of Meat, Fat, and Fiber Intake to the Risk of Colon Cancer in a Prospective Study of Women." *New England Journal of Medicine* 323:4 (December 13, 1990): 1664–1672.

World Health Organization. *Manual of the International Classification of Diseases, Injuries, and Causes of Death*. 9th ed. Geneva: World Health Organization, 1977.

Wright, Karen. "Breast Cancer: Two Steps To Understanding." *Science* 250:4988 (December 21, 1990): 1659.

Yu, Elena S. H. "Health of the Chinese Elderly in America." *Research on Aging* 8:1 (March 1986): 84–109.

Zopf, Paul E., Jr. *American Women in Poverty*. Westport, CT: Greenwood Press, 1989.

————. *America's Older Population*. Houston, TX: Cap and Gown Press, 1986.

Index

About the Author

PAUL E. ZOPF, JR. is Dana Professor of Sociology at Guilford College. An authority on demography, mortality, social theory and gerontology, he has published several texts on demography and is the author of *American Women in Poverty* (Greenwood, 1989).